18-50

RAF Wings over Florida

RAF Wings over Florida

Memories of World War II British Air Cadets

Will Largent

Edited by Tod Roberts

Purdue University Press
West Lafayette, Indiana

04   03   02   01   00   5   4   3   2   1

⊗ The paper used in this book meets the minimum requirements of the
American National Standard for Information Sciences, Permanence of Paper
for Printed Library Materials, ANSI Z39.48-1992.

Printed in the United States of America

Library of Congress Cataloging-in-Publication Data

Largent, Willard.
    RAF wings over Florida : memories of World War II
British air cadets / Will Largent.
        p. cm.
    Includes bibliographical references and index.
    ISBN 1-55753-203-6 (cloth : alk. paper)
        1. Largent, Willard.   2. World War, 1939–1945—Aerial operations, British.
    3. World War, 1939–1945—Aerial operations, American.   4. Riddle Field (Fla.)
    5. Carlstrom Field (Fla.)   6. World War, 1939–1945—Personal narratives,
    British.   7. Great Britain. Royal Air Force—Biography.   I. Title.

D786.L355   2000
940.54′4941′09759—dc21

                                                                99-059577

ᵔ᙮ Contents ᙮ᵔ

∿ Acknowledgments ∿

The help I received during the two years of gathering these accounts of Royal Air Force (RAF) cadets trained in Florida during World War II was nothing less than kind, understanding, and generous. Most of the material was submitted to me after lengthy correspondence with former RAF cadets who now live in many sections of the world. To them I am most grateful to have been entrusted with their recollections of the most important event of their lives. They made this book possible.

There were many persons of infinite patience available whenever I had a question. First to offer the key to tracking down British cadets who trained in Florida was Blaine Schultz, a former U.S. Army Air Corps cadet who was in Course 12 with British Royal Air Force cadets at Riddle Field, Clewiston, Florida. Memories of these RAF training days in Florida, and at other sites in the United States, are kept alive year after year in Great Britain. Riddle Field graduates have formed the 5BFTS (British flying training school) Association. Carlstrom Field graduates are bonded in the Arnold Scheme Register, which maintains information on British cadets who took their pilot training under the program developed jointly by Major General Henry H. (Hap) Arnold, Chief of the U.S. Army Air Corps, and the British Air Ministry.

Many helpful hands stretched out to me from Britain. John Potter, president, and Ray Searle, secretary, of the 5BFTS Association gave me a boost by mentioning in their group's publication my need for information. Norman Bate, MBE, registrar of the Arnold Scheme Register, was always quick to take to his word processor and dash off answers to my questions. Ben Travers, tireless researcher at the Royal Air Force Museum, always knew how to put his finger on obscure facts. Old newspaper clips were made available by Helen Riger, director of public relations at the prestigious Embry-Riddle Aeronautical University, Daytona Beach, Florida, the nation's oldest institution devoted to aviation training.

Ron Page, an extraordinary combination of talented writer, veteran pilot, and world traveler, was involved from the start with important data, facts, and figures. Vic Hewes, retired airlines captain and former RAF cadet, was unstinting in sharing his wealth of information, as were George Cordes, director of the Clewiston Royal Air Force Museum; Howard Melton, historian, Arcadia, Florida; and Bob Johnson of the Arcadia Rotary Club. The day-to-day activities at Riddle Field were extremely well documented by Jim Cousins, the highly respected former instructor and wing commander on the Embry-Riddle team. Beryl Bowden, longtime editor of the *Clewiston News* and founder of the Clewiston Royal Air Force Museum; Lois Blount, a former employee at Riddle Field; and Burnadetta Johnson of Arcadia offered their memories of the 1940s "welcome invasion" of the British. Old treasured copies of the Embry-Riddle publication *Fly Paper* proved of great value and were furnished by Eric Carlson, former instructor at both Carlstrom and Riddle fields.

Generous words of encouragement came from Gilbert S. Guinn, professor emeritus in history, Lander University, Greenwood, South Carolina, an authority on British aircrew training in the United States. Always coming up with the real "gen" when it came to interpreting British slang or RAFspeak was Alan Kelsey, former RAF sergeant air gunner and talented artist, now of Bradenton, Florida. Joe Terry, neighbor, friend, decorated World War II army veteran and history buff, called my attention to RAF Bomber Command Chief Arthur T. (Bomber) Harris's controversial policy of saturation bombing of both German military and civilian targets. Sarasota historian Jeff LaHurd was an authority on the "old town" near and dear to his heart.

I cannot give credit enough to my son-in-law, Tod Roberts, who edited, indexed, and "computerized" my original manuscript.

Lovingly last on my list of "thank you's" is the person who has always been number one with me, my wife, Gertrude. She has cheered me on, scolded me for loafing, and jumped in with encouragement when it was most needed. When I misplaced the dictionary, I could always call on her. What more can I say?

As to the errors in this book that readers will inevitably find, I, and I alone, am responsible.

Will Largent
Sarasota, Florida
April 16, 1995

Editor's Note

I am sorry to report that Will Largent passed away in Sarasota, Florida on January 21, 1998. As his son-in-law and editor, I have tried to carry out his wishes to publish this work. I thank Brooksie Bergen, Reed Clary, Bridget Dillard, Vic Hewes, Joe McCrary, John Potter, and Blaine Schultz for their help in publicizing the book. I also express my gratitude to the staff of Purdue University Press and Huron Valley Graphics for their valuable assistance in publishing the current edition of the work.

Tod Roberts
Sarasota, Florida
December 7, 1999

ᴄ: Introduction :ᴐ

A small news story in the *Sarasota Herald-Tribune* started it all:

> Memorial services will be held Monday, 10:00 A.M., at Oak Ridge Ceme-
> tery in Arcadia, Florida, for the twenty-three British Royal Air Force
> cadets who died during flight training at Arcadia and Clewiston in the
> 1940s. The fallen cadets were among the thousands of young British
> men who learned to fly in the United States during World War II.

My wife and I traveled the forty miles or so from our new home in
Sarasota to the old cattle town of Arcadia, where we followed signs
directing us to a narrow sand road. The winding path took us under
large oak trees to a back corner of the cemetery. A Union Jack was
flapping over a neat plot with twenty-three identical stone grave
markers in two orderly rows, twelve in one and eleven in the other.

We had arrived early on that Memorial Day morning, well ahead
of the service and (we learned later) an unusually large number of
persons—perhaps 600—to honor the fledgling airmen who lay in
the shadow of a bronze plaque etched with the Rupert Brooke poem
"The Soldier:"

> *If I should die, think only this of me:*
> *That there's some corner of a foreign field*
>
> *That is for ever England. There shall be*
> *In that rich earth a richer dust concealed;*
>
> *A dust whom England bore, shaped, made aware,*
> *Gave, once, her flowers to love, her ways to roam,*
>
> *A body of England's, breathing English air,*
> *Washed by the rivers, blest by suns of home.*
>
> *And think, this heart, all evil shed away,*
> *A pulse in the eternal mind, no less*

1

Gives somewhere back the thoughts by England given;
Her sights and sounds; dreams happy as her day;

And laughter, learnt of friends; and gentleness,
In hearts at peace, under an English heaven.

As we walked between the two rows, the leaves of the towering oaks rustled and stiff breezes whipped the ropes around the flag-pole in a rhythmic slapping sound. Nature was producing perfect counterpoint to a mournful yet stirring dirge that came to us on the early morning air. We turned toward the sound of the familiar bag-pipe tune "The Lament." About a hundred yards away a tall, kilted piper was practicing (we assumed for his contribution to the memorial program). Following "The Lament," he pumped out "The Minstrel Boy to the War Has Gone." We watched, listened, and gave thanks that so many other young cadets had been spared during those dangerous days in the Florida skies.

A thought kept coming back to me during the Memorial Day service that was attended by officers of the Royal Air Force, members of British-related organizations, World War II fliers, and throngs of service groups. Why not do a book in commemoration of those cadets who went through the training programs at Clewiston and Arcadia? This was the start of a project that brought me in touch with hundreds of men who looked back over five decades and gave me these stories of their youthful adventures in an atmosphere so different from that of their homelands.

Something personal also stirred me to capture these stories in print. I had flown as a U.S. Army Air Corps bomber crew member during World War II. As a radio operator/gunner, I had been on missions over North Africa, Italy, and Southern France. Quite frequently we were assigned British pilots based in Malta to give us additional fire power against German fighters as we carried out bombing strikes. It was always a cheering sight to look out from a gun position and see a British Spitfire or Hurricane flying with us, ready to take on the deadly German ME-109s or FW-190s. Who knows, I thought, one of those RAF pilots may have trained at Riddle Field or Carlstrom Field.

In this book I have limited my research to these two Florida training fields because it would take a multivolume effort to include reports from British cadets who trained at other sites during the war. Of course, this book makes no claim to be an official (or for that matter, unofficial) history of the various systems in place during the years 1941–45 to train British Royal Air Force pilots in the United States. Such a massive research project is best left to those historians accustomed to formidable bibliographies and armies of footnotes marching along in agate type at the bottom of each page. I say this with respect, admiration, and (I must confess) a bit of journalistic envy for those who have the stamina and tenacity to produce such scholarly and comprehensive works. My goal, quite simply, has been to relate selected stories about some of the men who learned the art of flying in a place far from their homelands and the slightly older American instructors who helped them earn their wings. Those whose adventures and memories I have reported were unusually gracious and generous in sharing their stories with me, looking back more than fifty years to recall experiences and events of their never-so-young-again days. These are the memories of British cadets, the American cadets who trained with them, the instructors who passed on their skills, and the Florida families who offered friendship, kindness, and encouragement.

The idea of training British pilots in the United States was President Franklin D. Roosevelt's response to Prime Minister Winston Churchill's radio address plea on February 9, 1941: "Give us the tools and we will finish the job," said Churchill, referring to war materiel and other support. Roosevelt's answer came in the form of the Lend-Lease Act, legislation that allowed Great Britain (and later, other Allied nations) to acquire war materiel against the promise to pay after the war was over. The actual plan had been hotly debated in the U.S. Congress over a two-month period, but the Lend-Lease bill became law in March, 1941. Now heralded as a major achievement in Roosevelt's presidency, the Lend-Lease Act may well have had the greatest impact of any single event in modern world history. Had it not passed, there is little doubt that Great Britain and the USSR would have lost World War II, with disastrous results almost certain to follow.

The Lend-Lease Act clearly spelled out where the United States stood, and it would involve an expenditure of more than $50 billion dollars between March, 1941 and September, 1946. United States and British military leaders huddled together to discuss scientific and intelligence matters. To the dismay of isolationists, the cloak of

America's "neutrality" became transparent. Military goods from U.S. factories were soon on their way across the Atlantic. As Lend-Lease convoys departed eastern U.S. seaports for Great Britain, they frequently were escorted by American warships that often encountered German U-boats in an undeclared shooting war.

Lend-Lease cleared the way for training British RAF pilots in many sections of the United States, including Carlstrom Field, Arcadia, Florida, and Riddle Field, Clewiston, Florida. The young cadets came to these bases by the thousands to learn the art of flying and to return home to defend their country. They also learned that being "strangers in a strange land" frequently provided an opportunity to revel in new experiences. The cadets forged new friendships that have endured for decades and over vast distances.

Some of the men whose stories I recount here say they can recapture the scent of orange blossoms when they glance through the log books they kept while flying their Stearmans and Harvards over Florida citrus groves. Many fondly remember the times when they buzzed over the homes of their Florida families to let them know to expect them for Sunday dinner. Memories of that short but exciting time in the 1940s are alive and well today. As the stories of these proud and hopeful aviators are told on these pages, some may even again hear the aircraft piloted by the young cadets over Florida.

Carlstrom Field, Arcadia, Florida

20th August, 1940
Winston makes his speech in the House [of Commons]. He deals admirably
with Somaliland and the blockade. He is not too boastful. He says, in refer-
ring to the R.A.F., "never in the history of human conflict has so much been
owed by so many to so few." It was a moderate and well-balanced speech. He
did not try to arouse enthusiasm but only to give guidance. He made a
curious reference to Russia's possible attack on Germany and spoke about our
"being mixed up with the United States," ending in a fine peroration about
Anglo-American cooperation "rolling on like the Mississippi."
Harold Nicolson, *The War Years: Diaries and Letters.*

Arcadia business interests, community groups, and ordinary citizens looked for many years after World War I to the skies above the
DeSoto County seat, hoping to see a rebirth of Carlstrom Field,
which had been a national air training jewel during World War I.
Named after famed flier Victor Carlstrom, who was killed in a 1917
plane crash, the field was where Eddie Rickenbacker won his wings
before becoming the legendary air ace with the Congressional Medal
of Honor. Determined to see the once-proud training field restored
to playing a key role in training U.S. Army Air Corps pilots, a group
named Airbase-for-Arcadia went into action during the 1930s. Born
with southern political savvy, they called in markers from state and
federal legislators who had benefited by support and contributions,
lean though this was during the Depression era. They urged editorial support from local newspapers, and while their approach contained a certain amount of puffery, the Arcadia boosters had plenty
of supporting facts to back up their boasts. The Florida News Service, a wire service serving many Florida newspapers, reported on
December 27, 1934:

> Proponents of the Airbase-for-Arcadia continue to be active in their
> efforts to secure a state-wide endorsement for the re-opening of the

federally-owned air fields in Arcadia that were used during the World War. Comprising a site approximately seventy miles long and thirty miles wide, this federal air base affords the greatest natural landing field in this state.

It is not the aim of the Arcadia air base proponents to take away from any other Florida city an air unit not in operation, nor do they oppose any other city's claims, but hope to secure for the state the location in Arcadia.

Important factors to consider in site selection are:

Past experience as a location for aircraft activities [Arcadia flourished as a World War I flight training site]; geographical location; topography; climatic conditions; transportation facilities, and (to cover remaining virtues) other requirements that such an institution may demand.

Concluding the story under a boldface subhead that read "Advantages of Arcadia Are Being Recognized by People All Over the State," the wire service reported:

Approximately 100 chambers of commerce, county boards, city councils, clubs, and other civic bodies throughout the state, as well as many influential private citizens have given their support to the movement. The project has the support of U.S. Senator Duncan U. Fletcher and Congressman William J. Sears.

There was plenty of political muscle wielded by Senator Fletcher and Representative Sears. Fletcher had been a two-term mayor of Jacksonville before serving as a Democrat to the U.S. Senate from 1909 until his death in 1936. Sears had been mayor of Kissimmee for four years and was elected as a Democrat to the House of Representatives, serving twenty years in split terms (1915–1929 and 1933–1937). With such an all-star cast of boosters, the Arcadia steamroller picked up momentum and more support as the selection process continued at creeping government pace over the next several years. The tip-off that Arcadia was getting close to winning the new air base came on December 4, 1940, when a team of Air Corps inspectors landed in a twin-engine bomber at weed-choked, snake-infested Carlstrom Field. It had been inactive for more than twenty years. On hand as self-appointed greeters were about one hundred Arcadians who joined local officials as a reception committee. All were hoping that the plane was bringing an early Christmas present, the long-awaited cash cow—the coveted air base. The inspection team was headed by General G. C. Brant, commander of Randolph Field, Texas, accompanied by two aides, Captains G. R. Storrie and W. J. Clinch. The purpose of the inspection was to determine a proposal to lease Carlstrom Field to a private non-government flying school to train pilots for the U.S. Army Air Corps. Acting spokesman for the inspection team, Captain Storrie, told an *Arcadian* newspaper reporter that the Air Corps started a program in July, 1939, geared to train 1,600 pilots a year. This figure was boosted in early 1940 to 7,000 a year and was divided among eighteen flying schools in the United States with facilities similar to what might come to Arcadia.

There would be no deal until a platoon of Army lawyers would review a three-foot stack of documents. The proposed training base would first call for leasing of Carlstrom to a flight training school. Next would be the acquisition of enough adjacent land for construction of hangars, barracks, mess halls, and recreational facilities. The entire complex would have to be large enough to handle the needs of air cadets taking ten-week primary training courses. The Army Air Corps also stressed that the base would be under the supervision of a qualified flying school contractor and capable of furnishing both ground and air instruction in the primary phase for military pilots. Before Uncle Sam signed on the dotted line, many related matters had to be worked out: housing facilities for school employees; construction of power and telephone lines; a water and sewerage system; and building of roads and auxiliary landing fields.

General Brant and the inspection team would meet later in the day in Miami with John P. Riddle, president of Embry-Riddle School of Aviation, who was interested in becoming the primary training contractor if Carlstrom Field was to see a new day. A team of prominent Arcadians worked closely with Embry-Riddle, a school of good record, to attempt to get community consensus on a plan that would bring dramatic changes and a population jump to the quiet community. The drum-beating Arcadian volunteers, guided by what was both patriotic zeal and the desire to see Arcadia prosper, were leading businessmen George T. Stonebraker, Paul Speer, Henry Avant, and Nate Reece Jr. They met frequently with Embry-Riddle officials on many sensitive matters.

Having never been repealed, the law of supply and demand was still on the books in Arcadia when word got out that the proposed school would bring in many new people. All would need living quarters. Whispers begat rumors, and rumors begat the real facts.

> They don't care how much rent comes to, it's all government money, so what the hell. A friend of my cousin said that she knew a woman in Montgomery (Alabama) who was getting fifty dollars for a room that she used to rent for twenty.

Some of those with space to rent (and who can blame them?) saw an opportunity to shake off some of the Depression blues by raising standard rental rates. To the credit of Arcadians, the dreams of rental wealth pretty much evaporated when *The Arcadian* issued these words of caution, December 19, 1940, in a news story that might well have been labeled an editorial:

> Naturally, the permanent personnel of such a school must be taken care of, at least temporarily, by means of facilities already at hand in the community; while it will no doubt mean that everybody will have to open up extra rooms and become landlords and landladies for a time, it is believed that the emergency can be met and everybody made comfortable.
>
> In this connection, those in charge appeal to the citizens to be reasonable in the matter of rentals. The people at the school will not be high-salaried people at all and must be able to find accommodations within their means. A tendency to be unreasonable in this respect might result in the decision to establish the school elsewhere, and Arcadia has waited too long for something of this sort to come along to have it driven away by the greed of a few people who might permit their anxiety over profits to warp their better judgment and sense of fair play.

Even though construction had started on the proposed field, making it highly unlikely that the project would go elsewhere, the newspaper warning in record-length sentences apparently did the trick. Then again, the realistic Arcadian knew full well that rental income was secondary to the genuine impact of increased spending by the newcomers, as well as job opportunities for local folks at a new military facility. Happy days *were* here again. The January 23, 1941, headline in *The Arcadian* read: "Carlstrom Field Long Dormant Is Galvanized Into Life Again As Buildings Rise on Old Site."

It was done. The long-awaited dream had come true and no balloon was going to burst. Happy New Year, indeed! There was no backing out now. Uncle Sam had reached down from Washington and patted Arcadia on its troubled head. Carlstrom Field was back in business. Now it was again doing what its slate-level land was born to do—teach flying.

The Arcadian story reported:

> Carlstrom Field is being born again. The famous field of the hectic days of 1917–18 will once more turn out fledgling pilots for the air service.
>
> The Riddle Aeronautical Institute plans an Air College that is the talk of the Army Air Service. The buildings are designed in the spirit of Old Williamsburg, which have been outstanding examples of the patriotic American architecture for generations. The architect, Stefan H. Zachar of Miami Beach, has endeavored to carry out the spirit of this old traditional style in designing these new buildings, which he feels is a happy fusion of the old and the new.

The buildings in the air field plan of the Riddle Aeronautical Institute were designed in a large circle and contained barracks, classrooms, dining hall, administration building, and recreation center. Architect Zacher said that the buildings were "designed to be in existence long after this emergency has passed." The design called for six tennis courts and a swimming pool, with a cabana and real beach sand imported from Sarasota, forty miles away. Flanking the recreational area would be one-story barracks buildings, each housing forty-eight students. Each building would be divided into twelve dormitory rooms with adjoining baths, each room housing four men. *The Arcadian* reported:

> These buildings . . . have their vertical axis north and south so that they may benefit from the cool breezes that blow during the warm summer months. The first building that will greet a visitor upon being admitted through the entrance gates of the school is the administration building . . . [it] will house all the administrative offices of both the school and army officers stationed there.
>
> The dining hall, with its up-to-date kitchen and storage facilities, is

located on the western side of the circle so that the prevailing easterly breezes will not tempt the young fliers, who will be in the classrooms on the eastern side, with the appetizing aromas that are so prevalent just before meal time.

The recreation building, which will be referred to as the canteen will be at the southern end of the recreational area and close to the hangars and flying field.

The school is designed to house and train two hundred and fifty students, but will be doubled most likely within the year to take care of five hundred. Two steel hangars will be built, which will be followed by four more of a new design (to minimize effects of high winds). Three 3-bedroom and two 2-bedroom houses of frame construction will be built on 60 × 100 foot lots. They will be used by school personnel and their families.

On Saturday, April 5, 1941, Carlstrom Field was rededicated and expected to give a repeat performance of what it had done during the war: training pilots to fly military aircraft.

Staff writer Bill Abbott of the *Tampa Tribune* covered the story:

In the atmosphere of a luxurious country club with the first 50 bright-eyed fledgling fliers standing proudly at attention, the most unusual defense project in the country was formally opened with ringing praise by army officers and local civilian leaders.

Out on the palmetto plain, eight miles from [Arcadia], the new air training center has risen in the remarkable speed of 60 days. Begun in February, it now is a self-contained city with two large hangars completed, with more than 20 planes in service and facilities for 150 flying cadets.

Other newspapers also gave heavy coverage to the event that Arcadians had been dreaming about for many months: *The Arcadian,* of course; the *Miami Herald;* the *Sarasota Herald-Tribune,* and the wire services.

Accepting the field for the army was General Walter R. Weaver of Maxwell Field, Alabama. MacDill Field, Tampa, was represented by Brigadier General Clarence Tinker and Col. Henry Young. Arcadians put on a first-rate show, welcoming the new field that was to pump thousands of dollars each week into a listless economy. But the

Aerial view of Carlstrom Field. (Photo courtesy of Victor Hewes)

PT-17 at Carlstrom Field, ready for another training flight. (Photo courtesy of Victor Hewes)

enthusiasm generated among locals for the rebirth of Carlstrom was mostly spurred by a Yankee Doodle/Johnny Reb love of the (then) 48 Stars and Stripes. It was an event—no, a celebration—that had not been seen since Armistice Day, November 11, 1918. The VFW and American Legion members, in their prime-of-life forties, marched straight and tall. The high school band members played their young hearts out, with the boys looking at the stalwart cadets with envy and the girls looking at the same cadets with a "wish-I-were-three-years-older" longing. The celebration was Norman Rockwell and John Wayne; small-town America and proud of it, by God; and Robert E. Lee—the small town that is caring and kind and sentimental, but always is quick to rise against injustice. "We can be real nahz," drawled a DeSoto County cowboy, "but if things ain't right we can be as tough as a fuckin' two-bit meat loaf." None of that French bubbly stuff to christen Carlstrom Field. Pretty Emma Marie Vance, Arcadia rodeo queen, did the honors with a bottle of genuine DeSoto County orange juice. General Weaver and General Tinker, both good sports, wore the cowboy hats presented to them as though they were the uniform of the day. Mayor Marshall Whidden issued a special proclamation calling on all Arcadians to participate in Carlstrom Field Week. Ed Wells arranged a typical Arcadia rodeo in honor of the distinguished visitors. Rodeo announcer George Stonebraker, decked out in his favorite gaudy orange cowboy shirt, was master of ceremonies at a program that lasted throughout the day and drew thousands of visitors to Carlstrom Field.

General Weaver praised the cooperation of local officials and citizens in seeing the new field become reality. "Carlstrom Field has inherited many of the fundamentals of the last war," he said, "and has received the backing of Arcadia and DeSoto County from the beginning. We can hope that this mutual good feeling will continue." Stonebraker pointed to the fine training record of the field during the war and said optimistically that the record would be exceeded this time. "We are going to see more training planes right here," said John Paul Riddle, "than at any other primary field in the country." [Stearman PT-17s would be used in primary training.]

The *Tampa Tribune* dedication story observed:

> Unlike the drab tents and barracks of other army posts, cadets at Carlstrom live in concrete cottages of Florida ranch-type architecture.

The mess hall at Carlstrom Field. (Photo courtesy of Bob Davies)

Barracks at Carlstrom Field. (Photo courtesy of Bob Davies)

Cadets on board the *Montcalm* bound for Canada, January, 1941. Victor Hewes stands third from the right. (Photo courtesy of Victor Hewes)

Cadets en route to Carlstrom Field from Canada, January, 1941. (Photo courtesy of Victor Hewes)

Their meals are served by white-coated waiters. They have a large swimming pool, tennis courts and other recreational facilities. Transplanted palms and bright beach umbrellas make the station a Florida showplace. The army is using its comfortable surroundings to bridge the gap between training planes and war planes to eventually turn out 1,000 primary-trained fliers every 10 weeks.

An Air Corps major told a fellow officer: "The theory is that if recreation is supplied right here, the student pilots won't seek it elsewhere. They can enjoy a swim and then get back to their studies." A three-stripe sergeant overheard the major's comment and muttered to a corporal, "Oh yeah? What bullshit is he handing out? How you gonna keep 'em here on the field after they've seen those eager little cuties in Arcadia and Sarasota? Hubba, hubba."

And so the droning of PT-17s over Arcadia became a welcome sound to the citizens, particularly the merchants, even though there were complaints about an occasional low-flying plane that startled cattle and caused a decrease in egg production. "Those hens don't know what to do with themselves," a housewife complained to a Carlstrom instructor from Buffalo who was renting one of her rooms. "I know what you mean," he said, "I ain't been home for three months and I hope my wife knows what she's doing with *herself*."

The ten weeks of primary training passed swiftly for the first class of U.S. Army Air Corps cadets; thirty of them graduated from primary training and were sent to basic training at Cochran Field in Macon, Georgia. No formal graduation exercises took place for the fledgling pilots. They were given a dinner dance and immediately whisked off to their new post as part of the Air Corps pilot training acceleration program. Only a few more American cadets were to be trained at the new Carlstrom Field.

The British Are Coming

The news broke May 31, 1941, that Carlstrom Field would receive ninety-nine British cadets for the ten-week primary training program starting in early June. They were part of the Arnold Plan, a program designed by General Henry H. (Hap) Arnold, Commanding General of the U.S. Army Air Corps, with the permission of President Franklin D. Roosevelt. The plan was offered in the face of the Neutrality Act, passed by the U.S. Congress in 1937 and designed to keep the United States out of other nations' wars. In brief, the plan called for British cadets to be trained in the United States as military pilots. Training in the United Kingdom was difficult because of unfavorable weather conditions, shortages of fuel and aircraft, blackouts, and bombing by the German Luftwaffe. There was little opposition to the plan in the United States—the overwhelming majority of Americans stood firm in opposition to Hitler's Nazi Germany, with only a small percentage holding out for an isolationist policy. In addition to Arcadia's Carlstrom Field, there were five other civilian contract primary flying schools for RAF cadets: Lakeland, Florida, Albany and Americus, Georgia, Camden, South Carolina, and Tuscaloosa, Alabama. Graduates of the primary phase at the civilian contract schools would go on to one of two U.S. Army Air Corps basic training fields—Cochran Field in Macon, Georgia, or Gunter Field in Montgomery, Alabama. There they would be expected to deal with the obstinate ways of the Vultee BT-13 before going on to advanced training. Advanced training, the final phase, also was given at Army Air Corps fields. Single-engine advanced training was conducted at Craig and Napier Fields in Alabama, using AT-6As. Cadets assigned multiengine training, using various twin-engine training aircraft, could be sent to Moody or Turner Fields in Georgia or to Maxwell Field, Alabama.

The first group of British cadets—the one hundred minus one—arrived to a rousing railroad station reception at Arcadia the morning of June 8, 1941. Throngs of Arcadians, including DeSoto County cowgirls in colorful rodeo regalia, greeted them. Tired, hot, hungry and sleepy, the young "hope-to-be-pilots" had been riding the train from Toronto for about forty-eight hours. They were dressed in wool civilian suits, neckties, and safari-style pith helmets. They were directed to Arcadia House, a hotel where employees (backed by women of the Trinity Methodist Church) served orange juice, tea, coffee, and doughnuts to the cadets on the hotel lawn. This was an appetizer!

A caravan of volunteers used their own cars to transport the cadets to the field. They settled into their quarters, showered, changed into tropical weight clothing and had breakfast. The British cadets would share quarters and some training with the fifty-three U.S. Air Corps cadets who had three to four weeks to go in the primary course before being shipped out to basic flight training. From this point on, Carlstrom would remain all-British until May, 1942, when the Arnold Plan at Carlstrom ended, and the field reverted to training U.S. Army Air Corps cadets.

The first Arnold Plan RAF cadets were in Class 42A, starting primary training June 9, 1941, on a course that ended August 15. Class 42A began with ninety-nine cadets ranging from age seventeen to early thirty. When the ten-week course ended, only thirty-six had passed a flight check and gone on to basic training, a 36 percent pass rate for the class. It should be noted that there were extenuating circumstances for the dismal elimination rate (64 percent) in the first Arnold Plan class: these cadets were the genuine pioneers in the Arnold training programs all over the United States. They had no opportunity to learn from those who had gone before them. The cadets were "personalities" in the same sense as we consider top athletes today. Their "different" speech patterns and mannerisms took America by storm. The newspapers couldn't get enough of them. They posed for endless pictures ("Shake hands with that elephant's trunk" at Ringling's Sarasota headquarters, or "Get close to the girls," at the Lido Beach Casino). There was no question that the newly acquired celebrity style of many cadets in Class 42A had a negative effect on their studies. Many of them became inattentive in ground school, partly due to their busy social lives (along with being regular churchgoers, Arcadians *did* know how to party). And there were the ground school instructors, all good fellows and most of them from Alabama, Georgia, Tennessee, and other southern states. "We had enough trouble understanding the Welsh and the Scots," a former cadet from London wrote. "Then we come to the States and encounter the southern drawl that we really like to hear, but just don't understand half of what is said. I never could comprehend what 'happy as a hog on ice' meant, or 'drunk as a skunk.' " (See "Watch Your Language," p. 245.)

There are no compelling explanations why the washout rate among Royal Air Force Arnold Plan cadets was twice that of the RAF cadets who went through one of the British Flying Training Schools (BFTS). Carlstrom Field (Arnold Plan) and Riddle Field (BFTS) were both staffed with American instructors and similar teaching outlines (BFTS, British, Arnold, American). There were some differences that could be of interest in technical journals. It would appear that the cadets in each program should have done equally well, but that was not the case. Total enrollment in both plans was almost identical; 7,000 for the BFTSs (washout rate of about 23 percent) and not quite 8,000 for the Arnold Plan schools (washout rate close to 45 percent). It was apparent early on that something had to be done to stop the flow of washouts in the Arnold schools. Late in 1941, Arnold cadets were given a four-week "familiarization" session at a U.S. Army air base before going into primary flight training. The program was designed to help them adjust to the new culture, climate, military discipline, and different foods. According to RAF officials, the last item, "adjustment to different foods," stunned some cadets, who were getting along quite nicely with unlimited American fare after living in England on the punishing one-egg-per-week, no butter, little meat diet brought on by wartime rationing.

But the effort to slash the alarming number of Arnold Plan washouts to a more acceptable level appeared to be too little, too late, and missed the core of what many felt was the real problem—the Arnold Plan itself, run under rigid U.S. Army Air Corps regulations. The Arnold Plan cadet was sent from one training post to the next (primary to basic to advanced) and never had more than ten weeks in one place before shipping out to another. This was in direct contrast to the British cadet in a BFTS program taking primary through advanced training in the same place. Stationed at one base for six months or more a cadet had plenty of time to learn the territory, make friends, and develop a feeling of security—all important to English boys who had never traveled far from their homeland.

Arnold Plan cadets had to suffer the West Point upperclassmen ("Chin up, mister, and shoulders back") hazing system that was part of the Army Air Corps training. Cadets were told that they must, under West Point rules adopted by the Air Corps, report to officers any cadets they saw cheating on an examination. The British code always held that it would be well beyond the pale to "grass" (inform) on a cadet who was breaking the rules (see Victor Hewes's comments, p. 41). Moreover, it would be *expected* that you would help him pass an examination. Tired of being ground down by the often silly, ever juvenile, occasionally sadistic, and sometimes plain stupid hazing incidents, RAF cadets took action, sometimes convincing base commanders to reduce, or, in some cases, end the old "tradition." Not a single letter from former RAF Arnold Plan cadets had a word of approval for the hazing system. "It eliminated more potential British and American pilots accidentally than the German Air Force did on purpose," one cadet said. "There was no reason for it at all."

The Arnold Register of England supplies these statistics of British cadets who entered training under the Arnold Plan in the United States: 7,885 started the six-month program; 4,377 graduated with both RAF and U.S. Army Air Corps Wings and diplomas; 81 died in training; one was reported as a deserter; 1,070 graduated as pilot officers; 3,307 graduated as flying sergeants; and 577 were retained as instructors at U.S. training bases.

"Taps" for Carlstrom Field

The headline in the June 15, 1945 issue of *Fly Paper* was bannered in all capital letters, two lines across an entire page: CARLSTROM FIELD WILL CLOSE ITS DOORS AFTER FOUR YEARS OF RECORD-BREAKING SERVICE. No byline was needed to tell that the story was written in the smooth professional style of Ralph Kiel, the flying school's director of public relations:

> Curtailment of Army Air Forces primary flying training makes it necessary to put Carlstrom Field on the retirement list. A veteran of World War I, Carlstrom becomes a veteran of World War II when the famous field closes its hangar doors and bids Godspeed to the last cadet on June 27, 1945.

The story pointed out:

> In more than four years of primary training, 7,500 pilot trainees (1,354 of them British) flew their PT-17 Stearmans more than 45 million miles with only one fatality, a record unsurpassed by any other training school at any time.

A one-column box in the feature piece was signed by H. Roscoe Brinton, general manager and legendary pilot, who said good-bye to "all Carlstromites," past and present. A small headline simply said, "And Now Farewell." Kiel's story covered all departments in giving credit to the success of Carlstrom. He then mentioned those persons who had contributed so much to the field on "the terrain of the Big Prairie," founder John Paul Riddle, of course, "who laid the foundation of this school that enjoys renown around the world."*

After praising parachute riggers "who dare not err," mess hall and canteen staffs "ever alert to the hunger of boys," and the citizens of the Arcadia area, Kiel closed with praise for the ground crew mechanics:

> A man might learn to fly by the seat of his pants, but it is the man on the ground with the wrench who makes it possible for that plane to take off and land.

The story closed with "Carlstrom's work is done."†

Profiles of Carlstrom Cadets

Bob Davies, A.F.C.

London, England

Bob Davies retired from the Royal Air Force after more than twenty years of service and embarked on a second career that helped satisfy his lifelong interest in classic automobiles. He took a job as driver for a multimillionaire Arab businessman. After four years, he was fired for stopping at the wrong entrance to Harrod's. He started driving the next day in London for Bahrain's ambassador and retired thirteen years, two Cadillacs, and three Buicks later. In his second retirement, Davies bought a two-year-old Chevrolet Caprice Classic Brougham and then a Cadillac de Ville. At last report he was zipping along to air and car museums in a Pontiac Firebird.

---

*Key figures in the Carlstrom Field years were: general manager Brinton; Len Povey, vice president in charge of flying operations; Lt. Col. E. G. Cooper, Army Air Force commanding officer; Major John Clonts, killed in action over France; Col. Stanley "Moose" Donovan; Lt. Col. George Ola; Major Clarence Porter; Major Sidney Netherly; Major William S. Hart; Lts. Alvin May, J. J. Graham, and Stanley Greenwood; Sgt. Eugene Busbee; Nate Reece, Jr., administrative assistant; Robert H. Davis; Jack Hunt; Andy Minichiello, and Joseph R. Horton.

†The state of Florida took over Carlstrom as a facility for the mentally ill and G. Pierce Wood Memorial Hospital, as it was known, received national accreditation in 1994, only to have it revoked seventeen months later because of neglect and mismanagement.

After I completed grammar school [high school], I entered an engineering college in London and enlisted in the Territorial Army [National Guard] during the Munich crisis of 1938. I was called up at the outbreak of war and served on a flak battery on the Thames Estuary until I was transferred to the RAF in late 1941.

My introduction to flying was on the Tiger Moth, and it was, to say the least, a disaster. It was quickly apparent that I—the last of my course to go solo—was not a natural pilot. If I recall correctly, it took about ten hours of instruction before I went up alone. Although I didn't know it at the time, I later learned that my flight commander was planning to take me for a "washout" check ride. For some strange reason (surprise, surprise) he sent me up solo and I escaped elimination from the course.

Fortunately I was in one of the last full elementary courses in Britain and I did about fifty hours of flying on Tiger Moths. Instead of going to advanced training in England, I was sent to Carlstrom Field, Arcadia, Florida. I can't remember how long it took me to solo the PT-17. I had to work hard at my flying, but when I left Carlstrom for basic training, I was confident that I would gain my wings. This illustrates the vital importance that Carlstrom—even with its appalling chop rate—played in my flying career.* I can remember clearly the time when three other RAF cadets and I met our very first American civilian flying instructor. He was Lloyd Whitney, a quiet-spoken man about thirty years old. It is to the everlasting credit of his personal leadership and dry humor that I was steered through those washout minefields unscathed.

So many memories of those Carlstrom days remain with me today—the first night in Sarasota when a few of us in our coarse blue uniforms went into what we thought was a rather elegant, sophisticated nightclub. We were trying our best to be nonchalant and worldly as we squirmed under the bright light that fell on us as we waited to be shown to our table. Cary Grants or David Nivens we weren't and our embarrassment must have showed. Seeing our plight, a dazzling black lady seated at a Hammond organ called out

---

*In using the term "appalling chop rate," Davies is referring to the high percentage of British cadets who failed to win wings as pilots. Early in the training program, the washout rate in Arnold Plan Courses topped 40 percent. Later, improvements were made in training procedures.

in a lovely throaty voice, "Come on in, boys, and make yourselves right at home. What can I play for you?" What a welcome! From that day forward I have loved Hammond organs!

Other memories also stand out—flying solo at one thousand feet above the thick Florida foliage and inhaling the heady scent of orange blossoms, the fantastic quality and quantity of the food, the quiet calm and reasonable attitudes of the flying instructors, the pathetic comic-opera personality of the American officer whose job it was to read to us the Orders for the Day after dinner. Lt. Kloppenstein would bluster and bumble his way through the dry-as-dust orders. All the cadets would wait until he finished and then in unison chorus something of a highly sarcastic nature.

Our evenings exploring the town and bars of Sarasota and the warm welcome we always received at the winter headquarters of the giant Ringling Brothers Circus were also great experiences. There was the time that five of us hired a Packard car on a twenty-four-hour pass and drove through the night to Miami—and then back again through the night to beat the midnight deadline. But this was not to be. We did not know that the big engine was gulping oil at such a rapid rate and we knocked out an end bearing some fifty miles from Carlstrom. A final memory was when, for the first time in my life, I really got drunk—more than a little squiffy! It all began thumbing a ride into Arcadia. A commercial traveler stopped his Buick and, after we got moving, said, "Reach around the back, son, and pull out my snake-bite remedy." I turned back and there was an unopened quart of Old Crow whisky.

"I keep it back there, son, in case a snake gets too close. Open it up and have a drink. I just saw a big rattler cross the road."

I cut the seal, opened the bottle, took a drink, and almost choked on the hot bourbon. I handed the bottle to my host and he took a long drink, taking it down like water.

"That's better," he said. "Uh-oh, there's another of those rattlers on the road. We'd better have another drink for protection." I made sure that the next drink I took was only a sip. It went down much smoother, as did the third and fourth, also sips.

Several hours later I woke up in a chair at his house (his wife, he had told me, "ran off with another man, a drinking fool"). When my head cleared, I looked about the house for my snakebite protector, but he was nowhere to be found. I heard a car engine in the drive

and there was the Buick, engine running, and the commercial traveler curled up asleep. I turned off the engine and saw that he was
breathing well. I walked into town and got a cab back to camp.

After the snakebite incident, I completed my primary training
phase at Carlstrom with results that laid the foundation of a most
satisfactory and entertaining career of almost 6,000 hours. Basic training at Gunter Field, Alabama, and advanced training at Craig Field,
Alabama, culminated in gaining my wings and commission on 5
September 1942 with Course 42H. I was posted to Instructor's
School at Maxwell Field, Alabama, and joined some thirty other RAF
junior officers as basic training instructors at Shaw Field, South Carolina. We had no senior RAF officer in charge and were under the
direct control of a U.S. Army Air Corps colonel. After ten months as
an instructor and over nine hundred hours in my log book, I received the sad news that I was to be posted back to the UK. From the
sunny, friendly skies of South Carolina, it was back to the blacked-
out and hostile skies of wartime England—a shocking change, indeed. To all of this had to be added the guilt that while I had been
swanning about [living it up] in the United States, my classmates of
42H had been operating in Bomber Command. Those boys had a
survival rate of somewhere between twelve and fourteen operations
[bombing missions]—not very good odds when a tour consisted of
thirty-two operations.

On my return to the UK I went through the usual "sausage machine" conversion to twin-engine aircraft and then four-engine Halifax bombers. I flew sixteen operations with 578 Squadron, 4 Group
Bomber Command. It was the policy of Bomber Command to fly
mainly at night. However, at this period of the war, General Montgomery told Bomber Command to temporarily go over to daylight
bombing of tactical targets. That brought me to 3 September 1944
when, at 17,500 feet, I was on the bomb run for Venlo Airport in
Holland. There were no apparent problems. Great weather, only
light flak, and so far as I could see, no German fighters. I recall my
bomb aimer saying, "bombs gone" just at the instant I heard an
almighty and mysterious crash. My first thought was that we had
taken a direct hit by flak. This opinion was shared by the top gunner: "There's a big fucking hole in the top of the fuselage about ten
feet aft of center!" Wait a minute! There was an equally large hole in
the *floor* of the fuselage just below the hole in the top, and the

chemical closet [toilet] and flare chute had disappeared. Then it dawned on us what had happened. An aircraft flying above us had dropped a bomb through our Halifax without exploding or knocking out controls. Strangest of all, not a crew member was injured when the aimless bomb ploughed its rough course toward land. A quick crew check revealed that the rear gunner was beginning to suffer from lack of oxygen. I told him to leave his turret and plug into the emergency bottle up forward. He declined my suggestion that he jump over the hole in the floor of the Halifax and take his crash position near the main spar.

Damage to our aircraft dictated that we initiate an SOS and set course for Woodbridge, a "crash" airfield on the coast, slowly losing height to 10,000 feet. As we approached Woodbridge, it was the decision of the entire crew to cancel landing there and go on to our Selby Airfield base in Yorkshire. (We were scheduled for leave the next day.) But deteriorating weather forced us to fly lower and lower, and at 800 feet with moderate turbulence and fifteen minutes from base, we were diverted to Old Buckenham, about ten miles east of Norwich. Back we went, virtually on a reciprocal course that turned out to be an anticlimax. The landing was fine, everything came down that should come down. We taxied to a halt and a very relieved crew went to debriefing and their respective messes for a much-needed glass or two.

As I was standing at the bar enjoying a whisky, I was approached by a young officer who asked if he could buy me a drink. He said that he was a bomb aimer on a diverted Lancaster and had been flying above our Halifax. He raised his glass to me and said, "Awfully sorry that I saw one of my bombs go through your Halifax. A terrible thing, really." What could I say in a situation like that? We shook hands, and that was that. I sometimes wonder if he remembers the incident, which could have gone so horribly wrong for both of us.

Why didn't the bomb go off when it hit my aircraft? The probable answer must lie in the fact that our bombs had a wind-driven propeller on the nose that had to make a number of revolutions before the bomb was fully armed and ready to explode on impact. This was a safety precaution that rendered a bomb harmless until it was well below its own aircraft. Obviously the Lancaster's bomb had not fallen far enough to be armed.

After completing sixteen operations with 578 Squadron, I was promoted to command "A" Flight No. 214 Squadron. The squadron had been flying Wellington and Stirling aircraft. If I had been put on Stirlings, I would promptly have gone LMF.* Few of us had confidence in the Stirling. The squadron motto of the 214th is "Avenging in the Shadows," or, as some crews would have it, "Lost, Frightened, and Fucking About in the Dark." Luck was with me. When I was posted to the 214th, it and a sister squadron were the only ones flying American aircraft—B-17s and B-24s—in Bomber Command. We did not carry any bombs, and the front armament and turret were removed. The nose section and bomb bay were filled with radio countermeasures for combating German night defense by jamming radio communications between German ground controllers and fighter pilots. The midsection was filled with WINDOW, short strips of metal foil (chaff) to be thrown out by the waist gunners as a means of misleading German radar used to target antiaircraft guns against the British bombers.

Davies and his aircrew took part in the highly controversial bombing of Dresden, Germany, on the nights of 13/14 February 1945. The U.S. Eighth Bomber Command struck the same target on the morning of 14 February, immediately following the British attack. The following day, the Eighth mounted another operation against Dresden. A total of 1,329 heavy bombers—804 British and 525 American—were thrown against the town on the Elbe River in that two-day period. Cries of outrage followed early reports giving the number of civilians killed as more than 130,000 in the fierce attack. The Dresden police chief reported the number killed at 25,000, and 30,000 "missing"—figures horrible to contemplate, of course, but comparable to other death tolls in similar ordeals, such as the bombing of Hamburg. Surprisingly enough, much of the fury was generated by Britons themselves, who had been on the receiving end of heavy bombardment when the Luftwaffe had the upper hand in the early stages of the war. It was as if a shocked Londoner might say, "Oh, yes, it is true that the Germans started the whole thing, but now it doesn't seem quite fair that we should take the same course they did. Not a level playing field, you know."

*Lack of Moral Fiber—the reason listed when a Royal Air Force crew member refused to fly any more operations.

What I remember of the Dresden raid is quite simple. The flight out was uneventful, and everything was clear over the target and also at our orbiting point about twenty miles south of the target. As I recall, we were briefed to orbit 2,000 feet above the bombing height of 20,000 feet. I was the first of the two squadron aircraft on radio and radar countermeasures circling and jamming for either five or ten minutes. Of course we had a grandstand seat watching Dresden burn.

When I first saw motion pictures of an atomic bomb explosion, it made me realize what the sight over Dresden that February night looked like. I had called out the navigator, bomb aimer, and the two wireless operators to see the spectacular column of smoke that rose in a mushroomed cloud higher than the height at which we had been circling. After the bombs had been dropped, we were wondering about our fuel supply. Was it enough to get back to base? We had flown all the way to the Czech border. I don't remember how long we had been flying when I saw lights on the ground and became uneasy. I told my navigator what was troubling me, the lights and all. "Lights," he said, "I should fucking hope so because we're over Switzerland."

The navigator discovered that the northwest forecast winds were about three times stronger than were given to us at briefing. I took my flight engineer's advice when he said, "Drop the rpms to a minimum, skip, and go for endurance." The last two hours were rather tense as we conserved every drop of fuel we could. I had already abandoned any idea of getting back to our base in Oulton, so I changed our destination to the emergency crash runway at Manston, putting the crew in ditching positions for the Channel crossing. We made a straight in landing at Manston and it was a very relieved crew that turned off the runway and taxied to dispersal. My engineer could not give me our fuel state, but he estimated that we had about ten to fifteen minutes left after landing. We returned to base and simply resumed our normal lives in the squadron. It was years later that I realized I had been a participant in what was reputed to be the "worst horror raid" of World War II.

Summing up my feelings about the Dresden bombing, I do recall that the intelligence officer said that he didn't know why we were going to Dresden, as the city had few industries and only a few optical factories. In addition, it was reported that the city was full of refugees from other towns. But to most of us it was just another operation for the log book. Like all the others it was another target, and I have no regrets that I was there. I have no feeling of guilt at all.

At war's end I had no wish to return to civilian life so I "signed on." I quickly converted to B-24s that were modified to carry passengers and bring back our troops from India. In a year the B-24s were scrapped in favor of Avro-York aircraft.

In mid-1948 all York Squadrons joined the Berlin Air Lift operation. After 420 round trips to Berlin, I was fortunate to be given the job of personal pilot to a three-star general with an Executive De Havilland Dove aircraft assigned for my use. My luck still held when the old boy got his fourth star and a new VIP aircraft, a Valetta. His new post put him in charge of all RAF activities in Germany. For two years we flew all over Europe. It was enormous fun for myself, my navigator, radio operator, and my two-cabin staff.

When the general retired I returned to the UK and to Transport Command, a small unit with the responsibility of flying and checking all transport crews once or twice a year. For the first time in my career, I was given a ground job training officer cadets. But even then, as all the staff were aircrew, we had several aircraft to keep us in flying practice. I was finally grounded at age forty-one and told that I would be retiring in two years. In London, I applied for a staff job at the Ministry of Defense and only regretted that the two years of "flying a desk" passed not swiftly enough.

I must say that I had very good innings during my years in the Royal Air Force. I want to emphasize that this was solely due to the environment and training given to me by my Embry-Riddle civilian flying instructors at Carlstrom Field in those days of early 1942. It is sad to know that Carlstrom Field has fallen on hard times because we all remember the old lady when she was in her pristine youth: green grass clipped to the point of being manicured, blue water in the spotless swimming pool, reflecting swaying palms, the buildings so white that they appeared sparkling at night.

But, then, I suppose we, too, have gone through some changes.

Diary of a Cadet

These excerpts from the diary of a nineteen-year-old British cadet cover a period from 22 February 1942, when he arrived at Carlstrom Field, Arcadia, through 1 May 1942. He graduated from primary training at Carlstrom and was sent to Cochran Field, Macon, Georgia, for basic training. The former cadet prefers that his name not be used.

22 February 1942

This morning we were issued our tropical kit. It is not too bad. We got two pairs of khaki trousers, three shirts, a belt, four pairs of underpants, and four vests [undershirts]. At 2000 hours we set off for Carlstrom Flying Field, Arcadia. We went by special train from Turner Field through the night and arrived at Arcadia the next morning.

23 February 1942

Arrived here at 0900. The place is very nice. White buildings set among green palms. There is a swimming pool and tennis courts. The barracks themselves are pretty good. Four men to a room, and each room has a very modern bathroom with a shower.

24 February 1942

We flew for the first time here today. The planes seem very easy to handle. I did quite a lot of turns and gliding and climbing and tried the controls out. Getting used to the ship generally. I did one landing, which seemed very easy.

1 March 1942

Flew again today. I taxied right out of the flight line to the taking off line and took off. Not bad for the first time. Did the same thing yesterday plus stalls. S-turns.

Undated Excerpts

Today it was very bumpy . . . very hard to keep the ship steady. Did a couple of spins. They are steeper and more alarming than in a Tiger Moth. Flew, did not do so well and got shouted at quite a lot, did takeoffs and landings. Had hair cut. Did two very good (or so I think) landings, two bloody awful ones, and one mediocre. Went into Sarasota at 1300 as tomorrow is my open post day. Booked a room at the Cypress Inn. It only cost me $1 and was very nice. Bought quart of Mr. Boston whiskey for $2.49 at Gator Liquors and carton of Camel cigarettes for $1.19 at B & B, took them back to Cypress Inn and opened Old Boston. It is not scotch.

Together with about a dozen other St. Andrews boys, I went round to all the nightclubs, including the Casa Madrid, and ended up at the Lido Beach Casino. Here most of us got very drunk and had a good time. They have an excellent band. Staggered back to the Cypress with Bob L. in a taxi. Arrived at about 0300. Got up at about 1030 feeling pretty bloody. Had a cold shower, which put me back on my feet again. We had breakfast at the Silver Coffee Shop. We walked round the town most of the day. It is very nice with masses of palm trees. I suffered from a hangover all day. I bought a camera for $15.50 and a case for $4.50. We caught the bus home at 1730. Stopped at Arcadia for something to eat and then went back to camp. Took what was left of Mr. Boston in with me . . . must not get caught.

No flying today. Booked a room at the John Ringling Hotel with Jack C. A crowd of us went to the Lido where we drank and generally made a nuisance of ourselves. At about 0300 H. E. got the worse for wear and I took him back in a taxi. B. C. came, too, and he and I went out for bacon and eggs. Got up at about 1000, feeling fine. Spent most of the morning on the pier taking snapshots. Went to Montgomery Ward store to look around for something for mother. Bought two pair of silk hosiery for $1.15 per pair. The clerk suggested one in "Canyon" and one in "Impulse." In the afternoon, we went to the Lido, where we swam in the sea and in the pool. It was very hot and the water was quite warm.

Back to flying . . . couldn't do a thing right and was thoroughly cursed. Charles, John, and Reggie did their first solo today. Today was even worse than yesterday. The instructor screamed louder and louder, and I got worse until he decided to bring us back and talk things over. We both got quite a lot off our chests. I have felt like death all day with worry as I am sure if I go on like this I shall be eliminated.

14 March 1942

A great day. Did fifty minutes dual in which I made thousands of mistakes and then I did my first solo. I felt great when Gus got out of the ship and said, "you can take her round yourself." I just nodded and set off. Everything went smoothly, and I sailed round the circuit better than I had ever done before. All I could think was, how marvelous to be all alone in the air at last. I did a good landing and the whole thing was miserably uneventful. Gus was quite enthusiastic (for him). He actually stuck his hand out and congratulated me. He sent me round again and then he got in and we went back to Carlstrom.

23 March 1942

Open post night. Went into Sarasota, booked a room at the John Ringling. Got up at 0900 and had breakfast. Walked three quarters of a mile out of town to get a driver's license. Came back and bought a pair of basketball boots to fly in. Bought swim trunks at the Tropic Shop for $1.49 (supposed to be $1.95), and then went to the Lido, swimming and sunbathing the rest of the day.

28 March 1942

Evans and Johnstone eliminated today . . . this makes six St. Andrews blokes.

29 March 1942

Flew all day, did six loops solo, also chandelles and Lazy 8s. New Yankee cadets arrived today.

30 March 1942

Had twenty-four-hour check this morning, got through quite easily. Did some power-on stalls, two steep turns, and a spin. Gus showed me a slow roll and a snap roll on the way back to Carlstrom. Wilson eliminated. This makes seven.

31 March 1942

Started ground school in the mornings, had a dual period in the afternoon. Couldn't do a thing right.

1 April 1942

Flew in the afternoon. Ken Barton eliminated today . . . this makes eight.

2 April 1942

Open post today. Had a long lie-in and had breakfast in the canteen. Played tennis with John in the afternoon. Maidstone eliminated . . . this makes nine.

5 April 1942

Paid $24 today.

6 April 1942

Flew in afternoon. Gus did three slow rolls without stopping the rolling action; he then did an Immelmann and a half roll. He did the same with Johnny and nearly made himself sick. He stopped flying for the day.

7 April 1942

Gus did not show up today. Had forty-hour check with Franz, did a chandelle, a Lazy 8, a spin, power on stalls, and forced landing . . . did not do anything well, but he let me through.

8 April 1942

Gus is back. Did a little surreptitious low flying today.

9 April 1942

Did snap and slow rolls for the first time solo. I could not do the latter at all. I managed to keep the nose up while I was upside down, but as soon as I started to come up the other side, it would fall off in a dive. Went into Arcadia at night with John, Fritz, and Bob to see *Design for Scandal.* Loved Rosalind Russell.

12 April 1942

Started aerobatics today, enjoyed thoroughly.

13 April 1942

Raffled my tennis racket amongst American cadets . . . quite a success, making $25.

14 April 1942

Had my first check with K. It only lasted for 25 minutes, but it put the fear of God into me. He looked just like a devil sitting there in the front cockpit and I certainly thought I had failed by the time he was through with me. However, I got through. Did pylon sights, chandelles, Lazy 8s, power on stalls, vertical reverses, snap rolls, slow rolls, and forced landing.

16 April 1942

Today I tried all sorts of aerobatics . . . a double snap roll, a snap roll from a vertical reverse, and a double loop from the recovery of a two-turn spin. Later on I joined the Short Snorters Club, which is a crazy kind of fliers club. To join you have to get two members to sign a dollar bill—your membership card. You have to pay them $1 each. Gus and the other instructor, Bob B., signed mine. You must always keep your dollar bill on you, as any Short Snorter can go up to another and ask him to produce his membership card. If he fails to produce it, he has to hand over a dollar. Gus asked all of us to have dinner with him tomorrow night.

17 April 1942

Had open post from 1330 today. At 1830 Denny, John, Johnny, and myself met Gus at HQ while Derek, Les, and Liggett met their instructor, Bob B. The four of us set out with Gus in his black Buick 8 and at Arcadia picked up Pilot Officer G. He turned out to be quite a decent sort of chap. A little way out of Arcadia, Gus stopped to buy half a dozen Coca-Colas with which we washed down a bottle of whiskey he had in the back of the car. After we had finished, we gave Gus the bottles, one by one, and he threw them out the window. When we got to Bradenton, Pilot Officer G. left us and we went into a very nice cafe to have dinner. Gus left most of his shyness under the influence of whiskey and everyone was very much at ease. Johnny behaved abominably the whole time . . . tried to give us the impression he was a terrific lady-killer and one hell of a man altogether. He got too familiar, and at one stage during the meal, he actually left the party to dance with some girl he had never seen before in his life. After an exceedingly good meal, we went over to the Tropical at

Sarasota, where we sat drinking more whiskey until about three in the morning. Nobody got the worse for wear except M., who kept going over to the Manhattan to look for some awful slag [coarse woman] of about forty whom he had dated, and we had a very interesting conversation. Gus seemed to be a little worried that he had to fly at 0800. M. did not come back with us, but stayed in Sarasota, and we were not sorry to lose him. We got back to Carlstrom at about 0430. Gus is one of the nicest guys I've ever met.

18 April 1942

Open post. Swam in afternoon, wrote to mother.

19 April 1942

Completed my sixty hours this afternoon, one period dual and one solo. We said good-bye to Gus for the last time today, a really great guy. I am really sorry to leave this place, although it is one step nearer the ultimate goal. I was feeling so rotten tonight that I went to the church service, which did me some good, I think.

22 April 1942

We were given leave from tonight [Wednesday] until Saturday night. Seven of us set off for Sarasota and tried to hire a car . . . no success.

23 April 1942

Tried again to get a car, and at noon we managed to get a very nice '41 Ford 8. It cost about forty-five bucks altogether, which is of course scandalous. We set off for Tampa, stopping at Bradenton for lunch. We put up at the DeSoto Hotel in Tampa. We had a very good meal at a cafeteria and then went out looking for night life. We wandered into a dance held by the Knights of Columbia [Columbus] but it was full of Catholic Fathers, who were friendly enough but who kept a sharp eye on us. They had no need to worry because the girls looked as if their sole ambition in life was to get into a convent. We left as soon as possible. While the others were at the dance, I had my first attempt at driving the car. Fritz came with me. Apart from once when I came round a corner onto the left instead of the right, I was more or less safe. When the others came out, I drove them over to the Chatterbox, a nightclub way out of town. The waitresses were on the whole extraordinarily good-looking, but apart from this and the fact that there was a very good Negro band, it was no different than any other night club. We stayed quite a long time. During the evening, I drove the car for a long time in the direction of St. Petersburg and back, managing to touch 70. It is a lovely car to drive. I drank nothing but highballs all night. At about 1300, V. and I drove Bob back to the hotel. When we got back to the Chatterbox, Tony Pastor's trumpeter (his brother) and pianist had turned up and were having a jam session with the rest of the band. It was terrific. Tony Pastor himself was there, pretty well drunk, and I got his autograph just for the sake of speaking to him. At about 0500 we had something to eat and went to bed.

24 April 1942

Got up at dinner time and had something to eat. We then went to hear Tony Pastor at the Park Theatre. Not bad, but his trumpet section was too loud. At about 0600 we set off for St. Petersburg, which we reached in about an hour. It looked like quite a nice place, but was absolutely dead at night. We spent most of the night at the Sundown Club.

25 April 1942

Drove back to Sarasota.

27 April 1942

Went to the graduation dance in Arcadia . . . it was a flop.

29 April 1942

We left Carlstrom for Cochran Field at Macon, Georgia.

30 April 1942

Arrived at Cochran early this morning. Right from the start we were given hell and made to double all over the place, getting equipment, etc. In the afternoon they gave us about two hours arms drill in the sun with the heat at about 90 degrees.

1 May 1942

A repetition of yesterday . . . wrote to mother.

Victor Hewes

East Point, Georgia

He took his first aircraft ride at age thirteen, and it was love at first flight. Since that day, Vic Hewes has logged 34,000 hours and 12 million miles. Hewes was born in Leicester, England, hard by Forest East on the Ratby Lane, future site of the Leicester Airdrome. Completed on July 1, 1935, the airdrome had its grand opening and young Vic Hewes had his grand aircraft ride. Hewes was then and there, 'til death do us part, committed to aviation as his life's work. He would pedal his bicycle the half-mile from his home to the airdrome to pick up odd jobs, "making a pest of myself" in exchange for an occasional free flight.

I was only seventeen when war broke out and my flying was curtailed, but I spent my time making model aircraft. Using identification charts, I made models of most of the new aircraft. The models were used by local stores as window displays for the war effort.

On my eighteenth birthday, I went into the Leicester recruiting office and signed up for the Royal Air Force. Some weeks later I was told to return to the recruiting office. I was put in charge of half a dozen other volunteers and was ordered to take them to Cardington near Bedford by train. We spent three days at RAF Cardington hunched over desks and taking one written exam after another. Then we would go to the medical building for tests and interviews that had no patience for personal modesty. One might say that everything was laid bare. At night we were allowed to go to the camp cinema. I still remember the pride that came over me being sworn in and given a lapel pin with RAFVR (Royal Air Force Volunteer Reserve) on it in large letters and told that I was accepted as a Pilot/UT (under training). At the Bedford Station, going home to await further orders, the Salvation Army ladies gave me a cup of tea, my first as a serviceman.

Soon after returning home (we had moved to Sileby), I was introduced to the commandant of Ratcliffe Airdrome, a Captain Mursell, who employed me as a second officer when I told him I was waiting call-up for pilot training. I worked in the control tower and flew as safety copilot on various operational aircraft, including Whitleys.

My main duty was to lower the landing gear in the event it would not extend. Ratcliffe was a great experience, a major learning step toward my later progress in the RAF.

Somewhere around the middle of 1941 I was called into service and reported to the Aircrew Reception Center at Lords Cricket Ground, London. The most vivid memory I have of Lords was the Free from Infection checkup held in the club room. Who could forget a whole line of chaps with their pants down being inspected by a medical officer using his swagger stick to manipulate any highly suspect genitals for a better view? We were kitted out [issued clothing] at St. John's Wood and learned to march in Regents Park. We never learned to enjoy the sorry food in the London Zoo cafeteria. Drill was not a problem because I had purchased a pair of service boots and wore them in [broke them in] before reporting to the reception center.

Our evenings were free and we went into the city, saluting everything in sight including doormen. We went to see the opening of a highly touted film, *Target for Tonight,* which had its desired effect by really pumping up the morale of us sprogs [raw recruits]. Looking back on the bit of celluloid claptrap today, one realizes what a piece of propaganda it really was.

When we completed work at the reception center, we were put on a train for Stratford-on-Avon and assigned to a large house on the edge of town on the river Avon, behind the church. Here we learned about King's Regulations [rules of service] and did our square bashing [marching drills] in front of the house. We studied Morse code in a building close to Shakespeare's house. Having detested the works of Mr. Shakespeare at school, I went out of my way to ignore the place. An office across the road from the church was where we studied aircraft recognition.* This was a piece of cake for me after my dedication to making models. At last we were no longer sprogs when we completed Initial Training Wing and were given our LAC [Leading Air Craftsman] badges to sew on our sleeves.

So it was back on the train in early October, 1941, for elementary flying training school at Cliffe Pypard near Swindon. We were at a temporary wartime airfield in the middle of the Wiltshire country-

*Aircraft recognition is a quick method of identifying model and "nationality" of an aircraft by seeing it in silhouette on a flash card that is shown for a second or two.

side, located on a country lane three miles from the main Swindon-Devizes road. Hitchhiking in Swindon was a major problem. Hitchhiking back to camp at night would be virtually impossible, so we spent most of our evenings in the NAAFI [similar to a U.S. Post Exchange, or PX] or in our barracks, typical standard wooden huts. The only permanent buildings were the hangars and the brick motor transport buildings. Those buildings and the gun butts [firing range] are the only things left standing on what is now once again a farm.

We began flying right away at elementary flying school. I was lucky to have a fine instructor named Sergeant Addy. He gave me four days of dual, a little over five hours of flying, and I became the first in my group to solo. We flew half days and had ground school during the other half. When weather prevented flying, we were sent out on our own on a route march. We chose to quick march down to the pub at Clyffe Pypard, make ourselves comfortable, and wait until it was time to march (slowly) back to the station. Then came a low blow! After forty-two hours I passed my training school proficiency check only to be told the next day that we would complete our flight training overseas. With my elementary training almost over, I tried my best to get sent to secondary flying training school in England, but without luck. We were loaded onto trains for Heaton Park in Manchester, to be posted somewhere away from Britain.

I remember someone, somewhere, asking where we would like to go: South Africa, Canada, or the United States. Most of us picked South Africa since we thought we would be near the fighting going on in the desert and could get into ops [combat operations] quickly. We were keen and eager, but rather oblivious to the distance between South Africa and the Montgomery/Rommel desert fighting. Canada as a second choice was acceptable, but few if any opted for the United States because we felt that training there would mean a delay in our being posted to combat duty. In true RAF fashion we ended up being sent to our last choice, the United States. There was nothing to do but wait at Heaton Park and try to forget about the frustration of delay in our desire for wings. I remember taking a WAAF to see *Dangerous Moonlight*, a movie about a British bomber squadron starring Anton Walbrook and Sally Gray. Michael Rennie, who would become an RAF flier, was in a minor role. Ironically, there was an air raid going on outside that drowned out the soundtrack of the noisy war action in the movie.

A few days later we were on the train to Scotland, where we boarded the good ship *Montcalm*, a soothing name. But we knew we were going to have a dicey trip. Why else would they hold a special church service for us when we sailed down the Clyde for parts unknown? Someone knew something that we didn't know, we thought. The *Montcalm* was an armed merchant ship with a hull full of timber. "You know lads," said a red-faced fortyish Scottish sailor, "we brought the timber along for your convenience. You can hang on to the wood if we get torpedoed somewhere along the way." The ship had a four-inch gun in the stern, but rode high in the water, causing her to have a perpetual dizzying roll. We were given hammocks to sling in the mess, but some of us had to sleep on the mess tables. This was not a good place to be because all of us were very seasick for the first few days of the two-week trip. We later learned we were being chased by U-boats on a regular routine.

We arrived at Halifax, Nova Scotia and watched Catalina flying boats doing circuits and splashes in the harbor while we waited to disembark. The train was waiting for us at dockside and snow fell on our way to Moncton, New Brunswick. The entire landscape was frozen with high snowdrifts, but the train, unlike those in England, was warm. The coaches were old sleeping cars and while the seats were not too comfortable, we could take the beds down. Even without mattresses we were able to get some sleep.

A second blessing was the food on the train. In a word, it was outstanding. It was wonderful to go to the dining car and be served on plates instead of having to line up with your irons [mess kits]. The RAF food and the alleged food on the boat were finally just a bad memory. When we arrived in Moncton we went at once to the first drug store we could find and ordered banana splits so huge that many of us got pretty sick. We were in Moncton for only two days, just time enough to get over the Atlantic crossing and turn in our gas masks before we were back on the train heading south. Two days after leaving Moncton, we pulled into a siding at Turner Field, Albany, Georgia.

There were many AT-7s, the twin-engine training plane, flying around. Our spirits were high as we all had hopes of going straight into secondary flying training and then on back home to be part of the combat war. It was quite a shock to learn that we were in fact at an American Initial Training Wing, a program almost identical to one that we had completed weeks before. We were treated as officer cadets, just like the Americans, even though most of us would graduate as noncommissioned sergeant pilots. The same thing applied to our navigators, bomb aimers, flight engineers, wireless operators, and gunners. Most of them were noncommissioned, although any aircrew member could be commissioned.*

Our cadet quarters were much better than the barracks of the Army Air Corps enlisted men and those that we had in the UK. We did not have to stack our biscuits [mattresses] and fold our sheets each day. Instead, we did it the American cadet way. We made up our beds with the blankets tight because you had to be able to bounce a coin on it. The shocker, though, was in the latrines. We were accustomed to stalls between the toilets. In the United States the toilets were lined up, arse to arse, all in a pretty row.

Shifting to a more pleasant subject, I remember the food at Turner as terrific, served cafeteria style—as much as one could eat. There was plenty of meat, ice cream, orange juice, fruit, and food that we had not seen for many months. We did not have to use or wash our own eating utensils or do cookhouse [KP] duty. In so many ways it was nice to be an officer cadet. The program at Turner was supposed to get us used to the American way of life.†

In retrospect, I now realize that the time spent at Turner was necessary and that the U.S. Army Air Corps did a pretty good job despite our opposition to a lot of petty rules and regulations. Most of us were in an "anti" mood at having to leave England. We had experienced the horrors of war. America had been at war for a scant two months and was still operating on a peacetime basis. We were on different wavelengths.

---

*Unlike the Royal Air Force, all World War II U.S. Army Air Corps pilots, bombardiers, and navigators were commissioned officers. Flight engineers, radio operator/gunners, and gunners were enlisted men, usually with the rank of staff sergeants or technical sergeants. There were exceptions: at one time there were pilot sergeants who were later commissioned, and some pilots who graduated from cadet training as flight officers, not commissioned, but above the enlisted rank. They also were later commissioned. On rare occasions there would be a fully qualified bombardier who was a noncommissioned officer. At least one technical sergeant was the lead bombardier of a Marauder Group in the Mediterranean Theatre of Operations.

†Although Hewes and his 42H classmates were scheduled to report to Carlstrom Field for primary training, they were posted first to Turner Field as part of a pre-primary indoctrination program. It was designed to cut down on the appalling percentage of washouts that had plagued earlier classes in the Arnold Plan.

American drill and customs did not come easy. We had to "pop to" and "get on the ball." An RAF cadet was called "mister," not "airman." Instead of doing an "about-turn" in drill, we had to "about-face." It was a whole new language, not anything like the Queen's English. Even commonplace words had different meanings that at times caused embarrassment to both Brits and Yanks. If you wanted to buy an eraser you did not ask for a "rubber." Americans used suspenders to hold up their pants. We used suspenders to hold up our socks, braces to hold up our trousers, not pants. Pants were underpants to us. (See the appendix, Watch Your Language, p. 245)

We had to get used to our relationship to the Negro community. Our American upperclassmen—many southern born—explained the segregation system to us in forceful terms. We should always sit in the front of the bus and we were made aware of those areas in town considered out-of-bounds. It didn't take long for us to understand that there was no love lost between northern Yankees and southern Rebels. The Civil War was apparently still being fought in the 1940s. We stayed out of those arguments.

The American custom of evening "retreat" to lower the flag involved a parade. Along with the American cadets, and led by the band, we marched to the flag pole, where there were "sound offs," "reports," and bugle blowing as the flag was lowered and carefully folded in a triangular shape while we stood in silent salute. The band usually played Sousa marches as the Americans called out their "hup, two, three, four" in cadence. We marched along singing a rather profane version of Sousa, driving the American officers crazy. They just didn't know that British Forces always sing when they march, and the words are far from the original ones.

In general, there was a lot of wasted time at Turner. We stood guard duty at the sewage farm [sewage treatment plant] all night, shot skeet, and took physical training during the day. We bought a revolver for five dollars and shot snakes at the Flint River. There were three movie houses in town, but little else except the red-light district. Some of the boys visited local families. One of these was the Carter family of nearby Plains, Georgia. "Miss Lillian" Carter, mother of president-to-be Jimmy Carter, was a most gracious hostess.

Then it came time to leave Turner Field and get on with our primary training. That was the word on 20 February 1942 when we arrived at Arcadia, a nice little town in central Florida, and home of

Carlstrom Field. The setting was like a country club with palm trees and a swimming pool. Buildings were all white stucco, with cadet living quarters up to par with any first-rate motel. There were only four cadets in each room, and we had our own bathroom. Again, the food was top rate. We could get a great bacon-and-egg sandwich in the little PX. We would sit and eat while listening to the latest Glen Miller records on the jukebox.

There was only one thing to spoil the fun: the West Point system. We simply could not get along with the hateful upper-class/lower-class hazing routine. The power of the senior class to control the lives of those below them was intolerable to the British, who were used to taking orders only from officers or noncommissioned officers. Under the West Point system, senior cadets could haze the lower-class cadets any time and, in fact, were encouraged to do so by their officers. When the first British pupils came over, the upperclassmen (U.S. cadets) would haze them by making them do push-ups, stand to attention in the so-called "brace" position, memorize and repeat on demand some silly bunches of bull [rigid and unnecessary military discipline] and, finally, eat "square meals" in the mess hall.*

After many protests were lodged against the West Point system, the practice eased up considerably for fear that a riot might result. In many cases the tactical officer or commandant of cadets was a West Pointer and a real s.o.b., as was the case at Carlstrom. Thoroughly hated was a Lt. Kloppenstein. Even the slightest infraction, such as a speck of dust at white glove inspection, would result in a demerit. Get five demerits and we had to walk around the flag pole for each demerit over five. We would have been happier doing something productive, even cleaning up the mess hall. With a monstrous war going on, the British considered the whole thing a glorious waste of manpower. The so-called West Point honor code prohibited lying and cheating, but then extended it beyond belief to British cadets. If we knew anyone who violated the code, we were to report them

---

*The "square meal," a favorite indignity, was heaped on a new cadet by a churlish upperclassman. The new arrival would eat each meal "squarely." He would cut off a piece of food about the size of a quarter and bring it to his mouth with his fork brought up from the plate and brought into the mouth at right angles. He would return the fork to the plate, moving it at right angles. Then he would place his hands in his lap and start all over again.

instantly, even if the miscreant was a close friend or, I suppose, your brother. This does not sit well with the British way of life. We believe in helping our friends. To illustrate a study in contrasts, I knew I would have difficulty in completing the minimum swimming requirement that called for each cadet to swim three lengths of the pool or fail the course. I knew I could not do this because of a swimming accident when I was in school, so I asked one of my friends to do the swim for me and turn in my name as though I had met the requirement. We helped each other in any way we could, and we enjoyed doing it because we were comrades in arms. Faced with the same situation, a U.S. cadet would have made a report of the "infraction" and his friend would have been thrown out.

I had to unlearn what I had been taught in England and did not get along with my first instructor, who did not want to send me solo even though, in my opinion, my flying was safe and more than satisfactory. I was lucky not to have been washed out for attitude when I asked for a new instructor. But I did get James Godette, who sent me solo right away. I never looked back and completed the Course in April, 1942, with an "above average."

Still in my memory bank are so many things at Carlstrom: Flying over the orange groves and smelling the blossoms even at one thousand feet, being forced to walk around the field with parachute on for some infraction of the rules, and checking the cockpit floor for snakes because one cadet was said to have bailed out when he saw one. Enjoyable indeed were our days off at Carlstrom. Four of us—Percy Mathews, Ian Harvey, Alan Huckle, and I—would rent a car in Arcadia for one dollar an hour and drive to Miami and West Palm Beach. The roads were straight and we drove with the gas pedal flat down whenever possible. It's a wonder that we survived those trips. The flying was safe, the driving was dangerous. Sometimes we would take a Trailways bus from Arcadia to Sarasota, where we were met by many residents who took us to their homes for the night. Not only were we wined and dined—we were taken to the Ringling Brothers and Barnum & Bailey Circus and the Ringling Museum. Percy Mathews and I were fortunate to meet a very nice family by the name of Mr. and Mrs. Lowell Morey, who invited us to their home anytime we could get away to Sarasota. We became good friends and corresponded for many years after the war. The folks in Arcadia gave us a dance and made us feel welcome.

It was quite a relief to graduate from Carlstrom. As was predicted, our class had a high washout rate, and many British cadets were back in Canada awaiting orders. But we had survived the primary course and were on our way to Gunter Field, Montgomery, Alabama, for basic training.*

We arrived at Turner Field on the Fourth of July. This is a holiday in the United States that celebrates the day the British got rid of the American colonies. On arrival in Toronto, the first thing we did was go downtown to Eaton's department store to be measured for our uniforms. They were ready for us the next day, and when we walked out of the store with those Royal Air Force wings on our pilot officer uniforms, we felt on top of the world. We stopped at the movies to see Bing Crosby in *Holiday Inn.* I will never forget that day.

Reporting to Maxwell Field, I went through the Instructor Training Course in record time and was sent to Cochran Field, Macon, Georgia, to instruct basic training on the BT-13 Vultees. I instructed one class of RAF cadets and then U.S. Army Air Corps cadets for the remainder of my assignment. One of the main attractions in Macon was Wesleyan College and its all-female enrollment. We visited the college for dates and often flew over the school for fun. The girls would sunbathe nude on the roof, and we would try to make a quiet glide approach before they could grab a towel or read the number on the plane. If some neighbor read a number and reported us to base, all hell would break loose when we landed. But the girls enjoyed the attention, and at times there were more Wesleyan girls in the Officer's Club at base than there were in college.

Flying over the United Methodist school formerly known as Georgia Female College was not the only infraction Hewes and his fellow instructors committed. They would fly into small local airports in Georgia and land, always attracting the attention of many local girls. After all, the local boys were away at war or in military training. The instructors would frequently take the girls for a quick flight around the field.

Hewes remained at Macon as instructor flight commander until June, 1942 and was then posted to western Canada to train Royal Canadian Air

*Hewes and his fellow classmates found no abuse of the upper-class/lower-class system from that point on in their training. Vic Hewes graduated from basic at Gunter and was sent back to Turner Field, this time for advanced training.

Force pilots. He and his wife, Betty (they met through mutual friends in Macon and were married there) lived in several western prairie towns and in Moose Jaw, Saskatchewan, until Hewes was posted to England.

After posting to an operational training unit, Hewes and his crew were told they had to get to India in a hurry. It took a frantic trip to London to pick up tropical uniforms and get inoculations, and then a breakneck ride to Poole in southeast England to be zipped away by Sunderland Flying Boat. They made quick stops in Tunisia and Cairo and traveled on to Iraq, Bahrain, and Karachi, Pakistan. Then the frenzied pace was over, and no explanation was ever given for the big rush to get them there. They idled away, sunbathing for twelve days in Karachi; fourteen days in Bombay, enjoying a complimentary membership at the Willingdon Sports Club; and five days in Calcutta, avoiding the mass of humanity.

From Calcutta they took a train to join 82 Squadron, based near Silchar in Assam, India, on the borders of Bhutan and Bangladesh. The squadron converted from the Vultee Vengeance to Mosquitoes. The airfield consisted of one strip cut into the jungle. Buildings were made of bamboo and palm leaves. Squadron 82 aircrews lived in tents and kept monkeys tied to the beds every night to guard against the large snake population. Deathly afraid of the snakes (for good reasons) the monkeys would not sleep and set off shrill chatters when a snake was present.

Chasing Japanese troops out of Burma was Squadron 82's main objective. The planes went after bridges, troop concentrations, airfields, and oil pipelines. Flights were over the Chin Hills, a rough mountain range offering no spot for a safe landing. In case of a bailout, there was little chance of survival. The terrain and punishing weather were far more dangerous than the Japanese troops.

Hewes writes: "Jungle animals were a problem. One Mosquito hit a water buffalo on takeoff and the crew burned to death in the flaming wreck. After that horror, a ground officer was assigned to drive a jeep down the runway and scare off the animals by firing his revolver. One night he withdrew his weapon from its holster and wounded himself almost to the point of ruining his love life forever."

At war's end, Hewes flew to the United States and was reunited with his wife and daughter. They went to England on the Queen Mary, and Hewes was soon back to civilian status working at his father's shoe factory. While on holiday in the United States with his family in 1948, Hewes was offered a pilot's job with Delta Air Lines and he jumped at the offer. At that time the company had only 185 pilots. Hewes flew thirty-four years with Delta, *retiring at the mandatory age of sixty in 1982. He founded a consultant company, Airport Safety Services International, and is a world authority on aircraft survival, including fire.*

Berkley Barron

Grand Turk, British West Indies

Berkley Barron, an aircraft pilot for about forty-five years, took a five-shilling hop in 1936 around Croydon Airport, England, in an Avro Tutor. That was his first time in the sky, and he was hooked. But he had to wait until 1941 to go on his first "real flight" as a Royal Air Force cadet at Carlstrom Field, Class 42F.

I was a British cadet in the Arnold Plan. I went through the normal U.S. Army Air Corps training. The United States, still at peace when we arrived at Arcadia, had insisted that British cadets in earlier classes not be seen in RAF uniform. They wore identical gray civilian suits to avoid, as one directive put it, a "uniform presence." It was a pathetic, feeble attempt to mask the fact that the United States was giving aid to a nation at war. What was the big secret? Young men with strange accents wearing similar suits and pith helmets (shades of Kipling!) were as inconspicuous as an Arcadian cowboy sporting a beret. By the time we were at Carlstrom, the chiefs in Washington and London realized that pretending America was not training British pilots to carry on the war against Germany was a ridiculous pipe dream. The order came down that our class would wear our RAF blues without any problem. "No more civvies," they told us. "Wear your uniforms and give the girls a thrill."

We followed the same "Mickey Mouse" guidelines and rules set down for new U.S. cadets. It was mostly nonsense and new to us. In fact, several in my class were eliminated because they had too many demerits for not obeying rules that didn't make sense to them. Some of the cadets who were eliminated went to Canada and continued their training there.* Great Britain had been at war for two years and desperately needed pilots. The United States, still in a position to pick

---

*There were several British cadets who washed out under U.S. Air Corps training and transferred to a British flying training school in the United States, where they won their pilot wings.

and choose, could enforce stringent rules that had nothing to do with flying performance. A high washout rate was not as alarming to the United States as it was to the British. But the majority of us took a "grin and bear it" attitude and after many discussions with the commandant of cadets, we managed to have some regulations relaxed a bit. White-glove room inspections and 45-degree blanket folds on your beds were required. Other "don't ask questions, just do it" regulations seemed minor indeed in light of global problems.*

While we were training at Carlstrom, the Japanese bombed Pearl Harbor, something I remember to this day, as do many others. I was in my barracks room listening to the radio when the "bombshell" went off, and everyone knew that the United States was in for an all-out war. Pearl Harbor was 6,000 miles or so from Arcadia and we couldn't understand why open posts [weekend passes] were canceled indefinitely. We thought this was utterly ridiculous in view of the fact that the Germans were only twenty-one miles from England. Finally the panic level eased, and we were again able to visit our friends in Sarasota.

It was a circuitous route that we took to get to Sarasota. We would telephone a car rental firm in Fort Myers to have a car delivered to Carlstrom. We would then take the driver back to Fort Myers and go on up to Sarasota. We would do the reverse on our return, a shining example of gasoline conservation! Sarasotans were simply fantastic to all of us. I became so attached to Mr. and Mrs. John Levinson and their daughter, Eleene, that I eventually returned to Sarasota with my family to live.

After graduation from primary training at Arcadia, I went on to basic at Cochran Field, Macon, Georgia. Following advanced training at Napier Field, Dothan, Alabama, I was awarded my wings (American). Strangely enough I never went through a ceremony to be presented with Royal Air Force wings and had to buy them in a military shop. I was commissioned and stayed on as an instructor at Cochran for nearly a year before going back to England. Because I had flown only single-engine aircraft, I wanted to fly Spitfires. After

---

*A former civilian American flight instructor notes: "The British kids had just come from a place where there was real war. They liked the flying and had no problem taking orders. But when it came to chicken-shit stuff, they wouldn't keep quiet."

completing the Royal Air Force advanced flying unit, I fully expected to be going to an operational training unit, the final step before hopping into that lovely Spitfire that was just waiting for me. This was to remain only a dream.

"Fighter pilots are no longer needed," the commanding officer explained in a pep talk at the base theater. "It is up to bomber crews to carry the war into Germany. I am calling for volunteers to go onto bombers," he said. He didn't get many and I never volunteered to fly bombers. No, I was "selected" the following day to be part of the Bomber Command. Also "selected" with me to pilot were many others: Brits, Aussies, and New Zealanders. "Not selected" for bomber duty were free French, Poles, Czechs, and a handful of Brits. They were to fly the coveted Spitfires.

So we were back in the training routine again. We started flying with an engine on each side and knowing what to do if one quit. Then we went on to four-engine aircraft and, finally, a finishing school on Lancasters before joining a squadron. In combat situations, our crew was very fortunate. We were attacked several times by German night fighters, but my gunners managed to knock down two and damage another. Once we were flying back-up to the Pathfinders, whose job was to circle the target and drop flares to keep it lighted for aircraft bomb drops. Our assignment was to throw out metal foil, which would create ghost images on enemy radar and, one prayed, would confuse antiaircraft gun crews.

"Are we the only Lancaster doing the back-up?" we asked ourselves as six of the dreaded and usually accurate 88-mm antiaircraft had us pegged. They followed us like hounds on a hare. Wherever or whenever we turned, the flak was right there, and we had a full load of bombs. The saving grace was that their range was off. The shells were detonating below us. But then they got it right. A shell burst in front of us at the proper altitude. A flak fragment ripped through the aluminum strip between the bomb-aimer's dome and the front turret and gashed my bomb aimer on the head. With blood streaming down his face (although the wound was superficial), he grabbed a first aid shell dressing hanging near him. His hands were slippery with blood and he couldn't grip the package to get it open. In pure desperation, the bomb aimer flung the kit to the flight engineer and yelled, "What shit! Can you open this bastard?" The flight engineer, understandably nervous, fumbled with the wrapping for what seemed to be five minutes, but really only a matter of seconds, and the bomb aimer called out, "Hurry up, I'm bleeding to death." While all of this was going on, my wireless operator was listening to "outside broadcasts," monitoring other aircraft, and then decided to change channels to our intercom. He was just in time to hear, "I'm bleeding to death." Thinking the voice was mine, he scrambled up, checked his parachute, and got ready to hit the silk if need be. Then clear thinking took over and calmness was restored. We returned home safely and years later got a lot of laughs about the operation when the German 88s had us pegged and we bounced all over Hamburg trying to dump those metallic strips. My wounded bomb aimer, a retired farmer in Wiltshire, will proudly show you his "wound stripe" for his head injuries. My wireless operator is a retired Australian school inspector who lives near Sydney. Ten years ago we had a reunion and only four of the seven crew members showed up. Now there are only three of us—I am the youngest at age seventy-one.

In 1952 I emigrated from England to the United States to a job with the United States Air Force as a civilian flight instructor in Columbia, Mississippi, and McAllen, Texas. When my contract ended after eight years, I went to work for the United States Army at Fort Rucker and did the same job except that the army didn't have jets. I then took a job in Sarasota and wore many hats. I was corporate pilot and personnel manager for the *Sarasota Herald-Tribune* and test pilot for Trans Florida Aviation and Cavalier Aircraft on P-51 Mustangs. As a DC-4 pilot I flew Northstars on a freight run to the Turks and Caicos Islands, British West Indies. I set up their internal airline schedule in the islands and am here now.

Allan Day

Uxbridge, England

Allan Day was in Class 42G at Carlstrom Field in November, 1941, and posted to Maxwell Field, Montgomery, Alabama, where he waited for flight training assignment.

We were given gray civilian suits because the United States was not yet at war. It was felt by the powers that be in Washington and London that it would be pushing the U.S. "neutrality" status a bit if foreign servicemen were observed being trained for war. Then, out of the blue, came Pearl Harbor and the United States was plunged into war. We were back in our blue Royal Air Force uniforms. No longer did we have to hide behind civilian clothes.

At Maxwell Field we were doing military drill, physical training, and all the other things that go into U.S. cadet instruction. We were introduced to the U.S. Air Corps methods of discipline, the honor system that required a cadet to report infractions of rules, and the hated demerit system that could result, for example, in losing a weekend pass. The long, rough ocean crossing, the confinement to camp, and the unfamiliar U.S. training made us feel somewhat dejected and even more anxious to be on our way to a flying field for training.

At last we were sent to Carlstrom Field in Arcadia, Florida. We were ready for our primary training segment at Carlstrom and would go on to other fields for basic and advanced training. All 224 of us were looking forward to doing our sixty hours flying time in the Stearman primary trainer (PT-17). We ranged in age from eighteen to mid-twenty. I was just over eighteen.

Our instructors were U.S. civilians with many hours of flying time. I was one of five pupils assigned to Ray Fahringer, who had done art work for Walt Disney Studio. This was evident in the excellent drawings in our flying handbooks. We were fortunate to be taught by this easy-going, friendly American. It was on Tuesday, 13 February 1941, that Ray Fahringer took me up into the warm Florida air and the mostly blue skies. What an ideal place to fly! I felt that I had truly "arrived." I worked hard at both flying and ground school lessons, navigation, and the like. We were helped in getting over some of the rough spots in training by a cadet in his early thirties, Flight Sergeant Benson. He was a British Army regimental sergeant major who had re-mustered to the RAF. With his years of experience, he guided us through many delicate differences we had with American ideas of discipline. Our high spirits did get out of control at times. Our routine called for us to be on parade each morning to raise the Stars and Stripes and each evening to lower the flag, with the bugler sounding the respective post. The ceremony was understandably a solemn one, and it simply wasn't following tradition when we broke out in laughter one morning as our eyes followed the ascending flag. There was a rather brisk wind, and we could not contain ourselves as the flag grew closer and closer to someone's underpants flapping at the top of the flag pole. Strange indeed that some American officers let it be known that they failed to see anything funny about the event and mumbled about handing out a few demerits. But Sergeant Benson came to the rescue with his ability to smooth troubled waters and all was well once again.

As we neared the halfway mark in primary training, we started aerobatics, a necessary part of combat flying that sometimes stimulates a bit of misplaced bravado in some pupils. One young cadet, a bit of a "wise guy," noticed what he thought was his instructor's failure to strap up his harness before going off to fly. (Actually, the instructor always strapped his harness, but he did it while the pupil was busy taxiing the aircraft out for takeoff.) Once the instructor and pupil were at proper altitude, the instructor demonstrated the proper way to execute a loop. Then, to the pupil's surprise, the instructor showed how to do a badly performed loop as the seemingly "unstrapped" pupil popped almost completely out of the cockpit but somehow managed to cling inside. Showing real skill, the instructor got the aircraft and cadet down safely—another tribute to Anglo-American ingenuity. The pupil seldom again tried to outsmart an instructor after such a lesson.

We would practice takeoffs and landings at an auxiliary field known as Myrtle. One day we were taking off as usual into the wind when all of a sudden an aircraft came toward us, trying to land downwind. How this wild cadet missed us I'll never know, but he had a complete inability to read a windsock and was eliminated from primary a little later.

Picture yourself in the situation of the cadet who rushed out one morning eager to do a few aerobatics, happy to be alive, ready to whip that little PT-17 around the sky! A perfect takeoff and you are all alone over Carlstrom Field, ready to take on those crack Hermann Goering yellow-spinner ME-109s and FW-190s. You are steely, tight-lipped, and grim as you scan the skies over Arcadia, looking up to the sun where the cunning Hun is hiding and ready to pounce on your Spitfire, which has logged fifteen enemy kills in only ten days. You cruise along, ever-alert, fearless, and ready for action. There is a mild sensation around your right leg, and you reach down and feel movement that isn't part of your body. You are stroking a writhing something or other about a couple of inches in circumference. It moves and slithers over to your left leg. You glance down and see a pair of hooded eyes and a flicking tongue. A snake at twelve o'clock low! What to do? The handbook says nothing about snakes. You are too brave to scream, but a few whimpers won't hurt anything. The cadet that experienced this frightening flight had a lot to talk about when he was finally able to take control of himself and bring his aircraft to safe landing. He didn't stop babbling until the medical officer gave him a few of what the Yanks called "courage pills," checked him over for any sign of snakebite, and pronounced him fit for duty. "You're O.K., son," the medical corps captain said. "It could have been worse . . . it could have been me." A ground crew made a thorough search of the Stearman for the unwelcome passenger, but the high-flying reptile was nowhere to be found. (He who slithers away will slither to fly another day.)

Welcome news in the form of a few free days came to us as we completed the Course a little ahead of time. A friend and I in our RAF blues (a sure ticket for a free ride) set off on the open road eastward toward Miami and Palm Beach. No sooner had we stuck out our thumbs when a huge car with a tall man behind the wheel and a passenger hunched in the back came along. The driver told us to get in the back, where there was "plenty of room." We quickly learned that the passenger, a short chap with expensive clothes, was the "boss" and when we told him where we were going he said to his driver, "take these boys to Palm Beach, George." Every now and then he would call out, "Pull over, George, I gotta get some fresh air." Then he would lurch out of the car and get sick into one of the wheel hubs. Between "getting fresh air" and rambling on about his various business endeavors, our host kept nipping at a flask that we declined to share despite his insistence. We finally arrived at Palm Beach and booked into a hotel for the night—amazed at all the stores and expensive displays of food, garments, and other items we hadn't seen in London for such a long time. The bright lights along the street made it appear that war was purely in our imaginations. When we left the make-believe atmosphere of Palm Beach for the more realistic setting of Arcadia, we had very little luck in hitchhiking. A truck driver finally gave us a lift to about sixty miles east of Arcadia where he turned off. We thought we were stranded as we walked along the dark road. We knocked on the door of a shack and asked a man if he would take us to our base. We explained that we knew there was a petrol shortage and were willing to pay for his trouble. That did the trick, even though we couldn't offer very much from our pay of twenty-two dollars every two weeks.

Looking back on those days, I remember responding to a posted memo stating that an American family would like to have a British cadet for the weekend. On Friday evening a lovely young lady came to the main gate to meet me. She took me to her home to meet her very nice folks, and I was given a comfortable room on the lower floor. Now here's where southern hospitality truly earns its reputation. After a wonderful breakfast (eggs, bacon, southern biscuits, and molasses), I asked where the young lady's father was. The man—he was probably about forty-five—heard me ask for him and shouted out, "Up here, son, come on upstairs." I found my way up the stairs and walked down the hall. Here he was sitting on the "throne" and reading a paper with the bathroom door open. I blushed with complete embarrassment, momentarily forgetting my question. "Just make yourself at home, son," he said. I remember mumbling some sort of thanks as I went out of the facility, closing the door as I left.

My log book at Carlstrom Field primary shows that I completed thirty hours solo and thirty hours dual flying time on 21 March 1942 and was on my way to Gunter Field, Alabama. It was good-bye to Ray Fahringer—what a wonderful guy! Then good-bye to Ian Turnbull, a roommate at Carlstrom, who told us tales of his commando training. He was armed with a bow and arrows and put into an inflated rubber dinghy. They would practice "invasion tactics," calling for Ian and his fellow Robin Hoods to steal up on German sentries and silently dispatch them to Nazi Nirvana. Ian decided that he wanted to be a pilot and so he re-mustered and joined the RAF.*

*Allan Day completed his basic course at Gunter Field, flying BT-13 Vultee Valiants, in May, 1942 and was posted to Craig Field, Alabama, for advanced training on AT-6s (Harvards). He was awarded British and American wings on 5 August 1942. Day scored second-highest in his gunnery class. Returning to England on a four-day crossing aboard the *Queen Mary,* Day did his preoperations training and was posted to a flight of Hurricane fighters. He completed a tour of operations from July, 1943 to September, 1944, based on the Scilly Islands and engaged in air/sea rescue operations with the Royal Navy and RAF high-speed rescue boats.

Henry A. Wright

Vienna, Austria

Henry Wright was in Class 42E at Carlstrom Field and arrived in Arcadia on November 6, 1941, one month before Pearl Harbor. The following is based on an article published in Arcadia's weekly newspaper, The Arcadian, *on November 6, 1986, forty-five years after Cadet Wright reported for flight training in Florida. Wright recalled those days and tells of the many friendships that he made.*

I was among the young men of Class 42E who came to Carlstrom Field, eager to become pilots. In these intervening years, I have continually recalled the friendliness and hospitality of the townspeople. I kept a diary at the time, but this is not needed to enhance the vividness of my memory. It has been useful simply to establish the precise date an event happened. This is how I know that it was 11 November, 1944, when a friend and I were wandering around town. A Mr. and Mrs. Scarborough from Zolfo Springs stopped their car and offered to take us for a ride round. You can imagine the alacrity with which we jumped at the offer—especially when we saw their pretty daughter, Loretta. We succeeded in getting a date with Loretta some days later and she brought a friend, Natalie, along with her. Evidently we could not sustain their interest and this proved to be the only date we had with them. About a week later, Bruce Davis, whose parents had orange and tangerine groves, invited my friend Len Woolgar and me to dinner. It was a terrific meal and good company. I often played table tennis in the church hall, where on at least one occasion Martha Scott from North Manatee Avenue played the piano. I sat her up on top of the piano—she was only about nine years old—and took a photograph of her. During the weekend of 29 November, a crowd of us hired a car and visited Sarasota. It was at a roller-skating rink that I met someone who I thought was the most beautiful girl in the world. I had fallen in love with Virginia Rhodes from 10th Street in Bradenton. I returned to camp with a burning desire to see Virginia the next weekend, but when Friday arrived I had no money. Then, as fate would have it, my hoped-for romance was shattered by the attack on Pearl Harbor. When that fateful day came, we were all restricted in travel and permitted to go no farther than Arcadia.

Bob Davies (third from left) with his Halifax bomber flight crew at RAF Riccal base in early 1944. (Photo courtesy of Bob Davies)

Members of the 42H class at Carlstrom Field in their prized aviator jackets. Bob Davies is at far right. (Photo courtesy of Bob Davies)

RAF cadet Bob Davies (right) poses with an American car. Great Britain had long been affected by gasoline rationing. (Photo courtesy of Bob Davies)

Victor Hewes in the cockpit of a BT-13. (Photo courtesy of Victor Hewes)

With our movements severely hampered, we became frequent visitors to Ray's Cafe—or was it a drug store? We became acquainted with Annie Stikeleather, the waitress. She was so charming and patient with all our bantering. I only hope that she is keeping her health and has found much happiness. Len Woolgar, a fellow cadet from my hometown of Croydon, Surrey, had been a reporter for the *Croydon Times.* It was therefore quite natural for us to visit *The Arcadian's* printing office. We saw the *Cadet's Handbook* being printed there and several Cropper printing machines.

We received 142 invitations for Christmas, but after tea on 23 December were told that eighty men would be required to stay on base Christmas Day. Our duty was to disperse aircraft within fifteen minutes of a warning as a precaution against sabotage. Volunteers were called, but so few stepped forward that names were drawn out of a hat. There was no grumbling over that arrangement. It was fair—we accepted it as our duty. But on Christmas Day, when we were told that *all* of us would be confined to camp, we became very angry. It made no sense to us, particularly because there would be no flying.

The most important person to me at that time was my flying instructor, Mr. Gordon Currier. He and his son, Sandy, drove out to the field on Christmas Day to give each of us some very nice cakes. When we had successfully completed our primary phase, Mr. Currier invited us to his home. I then moved from Carlstrom for basic training and after a few weeks had to write sad news to Mr. Currier. I had failed the course. His wife sent me a nice letter, commiserating with me in an attempt to raise my low spirits. Perhaps I had a guardian angel looking over me. I know that many cadets who became pilots did not survive the next few years.

I have been able to enjoy life since I left Arcadia, and I will never forget the people of Arcadia. I wonder where all those friends are now and whether the lovely Virginia had her own guardian angel to ensure we would never meet again.

Norman Bate, M.B.E.

Leicester, England

Norman Bate is registrar of the Arnold Scheme Register, a historical society of former Arnold cadets. The purpose is to register all persons who entered into, or were associated in some way with, the flying training plan for British cadets. It was named for the Chief of the U.S. Army Air Corps, General Henry H. (Hap) Arnold. The Register maintains contact with "Arnoldians" by means of The Arnold News, *a publication that assists in locating the almost 8,000 living or dead British student pilots who were enrolled in one of the Arnold Plan schools. (See Carlstrom Field chapter, p. 5.)*

I was twenty years old and six months when I reached Arcadia December 20, 1941. I had spent several weeks at Maxwell Field, Montgomery, Alabama, for orientation to American ways. We were given invitations to—and were expected to attend—functions in Montgomery homes. The move to Arcadia came as a bit of a surprise because there would be attractive girls in large cars parked outside the field, waiting to meet cadets and take them off for the day. This always seemed to be a little "off-putting" to young men like me who had never encountered anything like that in England, nor were we familiar with the "dating books" the girls kept. There seemed to be no security or real purpose in "dating" because we considered that almost as committed courtship, something that made us nervous. After all, so many of us were washing out that we couldn't bank on being in the Course the next day.

My first orientation flight took place with instructor Martin Gould, a very nice and capable young man, not to mention a master of the Stearman. Then I had Otto Van Schaich until I finished the primary course February 15, 1942. My first three solo flights had taken place on January 6, 7, and 8. There is nothing as exciting as that very first solo flight—flying an aircraft completely unaided. When I took a check flight with a U.S. Army captain, I really thought it was to be a washout flight. On the approach to land I saw all those aircraft on the ground crossing in front of my landing line. I angrily "gave her the gun" and shouted, "I'm going around again because there are too many crossing ahead!" It must have been the right decision because I never got washed out, and at my commission board several months later learned that my grade at primary had been well above average.

Instructor Otto Van Schaich was a heavy man, somewhat rosy-cheeked (as I was), with a mustache. He lived in a caravan [house trailer] with his wife. At the end of our Course he entertained us in the caravan prior to taking us to Sebring or Miami and gave us our choice of flying a seaplane or going to a nightclub. It was my vote that tipped it in favor of the nightclub where film star Fifi Dorsey was the featured act. It was a memorable night to close out our training at Carlstrom.

As we left Carlstrom for basic training, I knew that I had been a very lucky man. My room had been close by the swimming pool, the campus was absolutely superb, with palm trees, attractive buildings, entertainments, and an excellent dining hall. But, alas, the heat and humidity was something we had not anticipated. It was devastatingly cruel on the body. The field was exceptionally well run and we who survived the Course can count that primary time as really "making us" in terms of flying skills. We had a rude awakening at Cochran Field, Macon, Georgia, where we took basic training. The field was run by the U.S. Army Air Corps, unlike the civilian operation at Carlstrom. I commenced basic training with Air Corps Lt. R. T. Hall and I was convinced that I would soon be on my way to Canada as an "éliminé," with no chance at the pilot's badge.

Then a stroke of luck for me. Lt. Hall (my nemesis, I was sure) left to get married and Lt. Charles Thomas took over. He had been grounded for low flying below treetop level. In one instance he had been so low that he drove three high-ranking Army Air Corps officers in a merry Oldsmobile completely off the road. I was to become the first to accompany him on a repeat performance. His grounding had not changed his love of grass-high buzzing. We took off, gained altitude of 1,000 feet, and climbed no higher. Lt. Thomas, that carefree spirit, put the aircraft "on the deck"—down the railway tracks, so low that I was sure the mice were in danger. The low-flying "lootenant" was happiest when he had to dip the wings to avoid hitting small trees. It was a really hairy experience, a tutorial I shall never forget.

When our 42F Class reached basic the discipline became really and truly military and I guess that resentments did bubble up in our blood. The West Point system appeared to disturb the training concentration we'd come to expect. We wanted to concentrate on simply flying and learning essential subjects. The West Point hazing methods did strike us as being rather childish and bullying.

After basic training at Cochran, I was posted to Craig Field, Selma, Alabama, for advanced training and at the end of the course (that included gunnery training at one of Eglin Field's satellite fields off the Gulf Coast) I was given the "choice" of staying in the United States as a flying instructor. "If you want to stay on here as instructors," we were told, "you will go back to England and have your choice of any aircraft you want to fly." What bull, we thought, and so it was when we got back to England after six months of instructor duty in the United States. I returned to England in 1943 and was posted to an advanced flying unit at Peterborough, flying Miles Master single-engine aircraft. It was the first step in becoming familiar with climate and enemy intruders in the area by day or night. The words came back: "You will have your choice of aircraft." No one really gave me a choice!

In 1945 I relinquished my commission on grounds of ill health and was pensioned until 1952, when I was accepted into the Royal Auxiliary Air Force to take up a dual role of Meteor jet fighter pilot and squadron adjutant, flying on weekends and during our two-week summer camps.*

*Norman Bate was named Member of the British Empire by Queen Elizabeth at Buckingham Palace in 1993 for his work on the Arnold Register and Anglo-American relations related to Register activities.

Riddle Field,
Clewiston, Florida

Contrary to popular belief, the Everglades is not a vast swamp. It is a shallow river covering upwards of 5,000 square miles (12,950 square kilometers), measuring from fifty to seventy feet wide. It stretches from Lake Okeechobee in a southwest sweep of 100 miles, merging into saltwater marshes and mangrove swamps near the Gulf of Mexico.

As the British and U.S. cadets (ignoring "no flying zone" rules) flew over the "Glades," they were struck with the constantly changing kaleidoscopic views that make the Everglades what they are: one of the unusual—perhaps most unusual—of nature's montages. There are parts of the Everglades that have the look of a dense South American forest and there are the monotonous sections of vast open grasslands, relieved by the Caribbean "slash pine," the multicolored palmettos, and the mangrove trees, which grow as high as seventy feet. There are giant live oaks, banyan trees, and towering cypresses. Beneath the wings of Stearmans or Harvards were diamondback rattlesnakes and equally deadly water moccasins. There were black Florida bears and small brown deer, the prey of the panthers. There were otters, raccoons, and thousands of birds—white pelicans, egrets, and great white herons.

And there were alligators, crocodiles, and huge turtles in the shallows that the Indians called "Pahhayokee," the Grassy Water. The cadets were flying over the only Everglades on earth—wild, wonderful, awesome, and, at times, frightening.

The Royal Air Force wing commander, smartly turned out in his blue tropical wool uniform (tailored at Simpson's, Toronto), simply was not a person to be easily shocked. At age thirty-two, he was a decorated fighter pilot, holding the Distinguished Flying Cross.* He was credited with five enemy aircraft destroyed during the Battle of Britain, and on a number of occasions he himself had been shot

*The Distinguished Flying Cross (D.F.C.) is the military citation awarded for "heroism and extraordinary achievement in aerial flight."

61

down. But Wing Commander Harry Hogan, with eleven years in the Royal Air Force, admitted many times later that he "was taken aback, *really* taken aback," when a civilian pilot at the controls of a Stinson airplane he was passenger in made a surprise landing in Florida's Everglades country.

Hogan must have told this story dozens of times over the years: "There was nothing but grass down there and, other than that, miles of nothing. This chap leaned over and said, 'I'm going to shut down the engine and I predict that underneath where the propeller stops is where we will build the new British flying training school." The pilot was John Paul Riddle, thirty-nine, who had already attained legendary status. Riddle was operating Carlstrom Field at Arcadia for the purpose of primary training for British and American cadets and now had been authorized to build a British flying training school at Clewiston, Florida, a fact unknown to Hogan at the time. When the Stinson rolled to a stop in the rough grass and we got out, Riddle extended his arms and said, "Right where we are standing will be the runways." He pointed eastward. "That's where our control tower will be." Looking to the north, Riddle said, "The dormitories will be over there and the mess hall alongside of them."

The spot where Riddle had set the Stinson down early in 1941 with Hogan had, in fact, already been selected by Riddle and RAF Group Captain D. V. Carnegie as second choice to a site in Sebring. Sebring was ruled out when the U.S. Army Air Corps claimed it as a possible airfield in the Panama Defense Plan. Hogan was based at Maxwell Field, Alabama, and was merely checking out the general location of all fields where British cadets would be trained. The RAF brass had assigned him to look after the interests of British pilot cadets in the United States. (Hogan's abilities were so highly regarded that he was awarded America's Legion of Merit. He retired as RAF Air Vice-Marshal.) Beryl Bowden, former *Clewiston News* editor, said in a 1993 interview that Riddle told her in 1943 about how he had "pulled one over" on Hogan. "John really enjoyed his practical joke," Mrs. Bowden said. "And, really, it was unusual for him because he was always quite a serious man and I know that he liked Harry Hogan very much."

The official announcement of Clewiston as the proposed site of the RAF flying school came slowly, with one rumor following another: the airfield was going to Sarasota or some place in Manatee County or some place in Georgia. There was no huge ground swell in Clewiston for the air field, no beating of business drums as there was in Arcadia. But that didn't mean that a proposed training school site would get short shrift. No, residents of Clewiston clearly indicated they would welcome the British cadets. That Anglo-Saxon-Celt strain was always strong in the "Sweetest Town in the World," so named because Clewiston was, and is, corporate headquarters of the giant U.S. Sugar Corporation. A great number of Clewiston surnames could well have been a passenger list of early British immigrants.

When the *Clewiston News* hit the streets July 11, 1941, a page-one story fairly well confirmed that a British flying training school would be located in Clewiston. Scooping the large city newspapers, the lively weekly carried the headline:

TRAINING FIELD IS VIRTUALLY ASSURED ON CLEWISTON SITE

A headline deck was more positive: "Riddle Training School For British Cadets Is Planned Here." Hedging a bit by stating "no official word has come from Washington," the *News* labeled it "virtually certain that a flying school will be established to train British fliers on a site eight miles west of Clewiston." The story said that the school would be operated under contract with the British government. "The site chosen," the *News* reported, "is four sections of land lying to the south of and adjacent to the highway. The northwest corner of the tract is located where the new bridge has been built across the canal to the highway." The story pointed out that Riddle Field was to be similar in type and construction to Carlstrom Field, where the first group of Arnold Plan British cadets were completing primary training. Still leading the newspaper pack, the *News* reported the following week that Riddle Field construction got underway on Thursday, July 17, 1941. "A bulldozer was moved to the location and work began on leveling of the bank of earth along the highway canal," the reporter noted. At the time the weekly paper broke through the wall of governmental secrecy, John Paul Riddle took a careful position on any specific details concerning the field. "Plans for government contracts must remain secret until the time is proper for their disclosure," Riddle said. But the alert *News* followed through on Riddle's statement by reporting: "It is known that the school is a civilian flying school that is intended to be used as a preliminary flying school for the training of young British cadets in the United States." It was apparent early on that the newspaper was tapping a leak in high places, as it regularly beat the dailies at their own game.

Chosen as contractor for the Riddle Field project was the C. F. Wheeler firm of Miami, with C. W. McSheehan acting as project manager. Eager to see the construction completed in record time, Riddle insisted on something else in the contract: the project manager would be required to live in Clewiston until the last coat of paint had been applied. And dried. McSheehan and his wife took an apartment at the Clewiston Inn.

Acting as self-appointed motivator of the project, lanky, tousle-haired John Paul Riddle would appear unexpectedly at the job site from time to time. Dressed in the seersucker suits he favored, Riddle would scramble around building materials or bins of greasy plumbing fittings, seemingly unconcerned about damage to his apparel. "The quicker we get this job done, the faster we can train those British kids to fly," was his favorite phrase, and he used it so often that workers would scurry for cover whenever he came in view.

Riddle's pushing and project manager McSheehan's engineering knowledge paid off. By mid-August, construction had been started on almost every building. One barrack had been completed. Plans called for all buildings to be constructed squarely in the center of the four square mile tract occupied by the field. Layout of the complex was in the form of a huge diamond, with the ends of the diamond pointing north and south. With the exception of administrative headquarters, buildings were located in the south end of the diamond. Space was earmarked at the north end for any buildings that would be needed in the future. A large steel hangar to house about thirty aircraft would be erected at the southwest end of the diamond, with the possibility of another hangar going up if enrollments increased. Concrete block walls were up on one large barrack and partially up on two other large barracks. The foundation for the recreation hall was poured and work on the mess hall was half-complete.

The two main barracks were each 280 feet long and 26 feet wide. They contained 18 rooms measuring 12 feet by 25 feet. Smaller barracks had 10 rooms of the same size, with each room in the barrack housing four cadets. All buildings were of concrete block construction on reinforced concrete wall foundations. Roofs were made of heavy slate over felt. Exteriors of all buildings were white stucco. All interior walls were plaster ones.

Largest of the buildings was the mess hall, 130 feet by 130 feet. The classroom building for ground training was 25 feet by 130 feet; an infirmary was approximately the same size. South of the barracks was a spacious recreation hall in a U-shape, measuring 59 feet by 89 feet. Adjacent to the recreation building was the flight operations building, 36 feet by 22 feet.

Between the two large barracks were outdoor recreational facilities: tennis courts, a basketball court, and volleyball courts. A swimming pool, 31 by 22 feet, was located between the two smaller barracks. There were two artificial lakes at either side of the complex. Housing facilities at the field were for cadets only. There were about 100 instructors, mechanics, and other technicians who were housed in private facilities away from the field.

Rooms for rent were, in the wonderfully descriptive southern style, as "scarce as hen's teeth." Instructors and ground staff were paying anywhere from twenty to sixty dollars a month for single rooms. The handful of houses for rent were quickly snapped up by those who arrived on the scene well in advance of the Riddle main crew. There were no rooms at the Clewiston Inn that hadn't been taken at the year-round daily rate of three dollars for room and bath. The nearby Moore Haven Hotel, with a lower rate, hung out a "no rooms available" sign early in the mad rush for lodging.

Getting the 5BFTS facility into operation didn't involve just bricks and mortar. All other components going into the showplace facility were shining jewels in Britain's RAF training program. But a sparkling new airfield with all the amenities (even an ice cream/soda bar) could not turn raw cadets into pilots. The ability of these cadets to go against Germany's finest, the Luftwaffe, had to come from their top-notch instructors, the most important ingredient.

So John Paul Riddle, confident that the construction phase of the field was on track and in fact ahead of the most optimistic estimated completion date, turned his hand to staffing the field. Riddle determined that logistics called for him to draw senior instructors from Carlstrom Field over to Clewiston, in order to establish a nucleus of experienced instructors at the new field. The highly respected John Cockrill, a senior instructor who had been with Riddle for a good number of years, was the point man in sounding out Carlstrom instructors who might be interested in going over to Riddle Field. John Paul Riddle was not about to do damage to his school at Carlstrom. His plan was a sound one, calling for moving senior instructors from Carlstrom over to Riddle Field to teach the basic and advanced courses called for in the British curriculum, and getting instructors completely qualified to instruct primary training to fill in the gaps at Carlstrom.

When the offer was made to the Arcadia-based instructors there were only about fifteen takers out of more than one hundred men. Most of them were already established in Arcadia from a housing standpoint. Some of them had wives who had become involved in community affairs and felt at home in the friendly community. There would be no big hike in salary, and living costs were lower in Arcadia than in Clewiston. Finally, who would want to go from instructing on the PT-17, a sweet-flying airplane, over to the Vultee BT-13, a basic trainer that was feared more than respected by most instructors?*

Heavy rains had cut into flying time at Carlstrom and the first Courses of cadets at Riddle were told that lost air time would have to be made up on weekends. Cadets were ordinarily off from noon Saturday to 9:00 P.M. Sunday. Their day began at 5:00 A.M. on the drill field. Ground instruction or flying time took up much of the day and lights were out at 9:30 P.M. There were about sixty aircraft assigned to the field: twenty PT-17s (Stearmans); twenty Vultee BT-13s, and twenty AT-6As (Harvards to the British, Texans to Americans).

The time that it took to construct the Riddle Field training facility (at $2 million) may well stand today as a record for fast-track engineering. The first backhoe hit Riddle Field soil July 17, 1941. The first contingent of RAF cadets, eighty-nine of them, moved into barracks on September 25, just two months after ground was broken. The boys arrived mid-morning, were assigned their quarters and given an hour's free time before lunch at the new mess hall (menu: pork

---

*One seasoned RAF pilot had this comment on the BT-13: "It has a tendency to snap roll on landing, which can be rather disconcerting. Other than that it isn't very dangerous." (After several Courses had gone through primary on the PT-17 and basic on the BT-13, the BT-13 was eliminated; cadets then went directly from primary to AT-6s for both basic and advanced.)

roast, yams, green beans, and choice of apple pie or ice cream—or both, the choice of most cadets).

Even John Paul Riddle was impressed with the performance of the contractors. When he arrived at the field to give it a quick look, a worker on a barrack roof called out a warm greeting. "Good morning, Mr. Riddle." Then the young roofer tossed out a version of the shopworn phrase made famous by "the boss": "I have everyone on my crew working at top speed," the roofer said. "You know, Mr. Riddle, I just keep tellin' them that the quicker we get this job done, the faster you'll be gettin' those RAF kids in the air."

"That's the spirit," Riddle said, "I never thought of saying anything like that myself. Seems to work, though."

The first group of cadets actually were members of 5BFTS (Riddle Field, Course No. 1 and Course No. 2) and had received part of their training at Carlstrom Field while waiting for assignment at Riddle. The first cadets to start complete training at Riddle were fifty students from Course 3, who arrived October 8, 1941. All were undergraduates of Oxford, Cambridge, or Edinburgh universities.

Between July 17, 1941, and August 25, 1945, about 1,800 Royal Air Force cadets (Courses 1 through 24) entered the six-month pilot training at Riddle. Graduating and winning wings as pilots were about 1,400 cadets. There were 300 graduates commissioned pilot officers and 1,100 who were given pilot sergeant rank. A sizable number of the pilot sergeants were later commissioned, some rising well above mid-level RAF ranks. Those students who washed out were generally sent to Canada for record evaluation and placement. A small number were put back into pilot training, and in some cases did very well. Others who did not win pilot wings were trained as bomb aimers, navigators, wireless operators, or gunners, and often served with distinction on many fronts.

When the long-awaited surrender of Germany was announced May 8, 1945, there was intense speculation about what was in store for the 101 cadets in Course 25. They had arrived at Riddle on April 3 and the 100 cadets of Course 26 were due to arrive on June 19. Opinions were divided as to whether training would continue or be canceled. Would there be an increased need for RAF support in the Pacific Theater? It was anyone's guess as to how long the war would last. The high-level decision was made swiftly by RAF brass: Course 24, in training since January 22, 1945, would be permitted to graduate, as scheduled, on August 25, 1945. Word came from London that all future training would cease because of "an oversupply" of pilots; indeed, a number of them had been shifted in 1944 to British Army ranks and trained as glider pilots. Superfluous pilots? A thought that would never have crossed one's mind a few years earlier.

The 201 cadets of Courses 25 and 26 got a mere "incomplete" mention on the 5BFTS log. "No drums, no glory," someone scribbled on the bulletin board notice in the Intelligence Room. Underneath that sentiment was another penciled freedom of expression: "And no sodding flak or sodding Jerries."

Awaiting orders to return to Canada for reassignment, the "incomplete" cadets had little to do to occupy their time. One highlight came on Polling Day, July 5, 1945. The cadets, along with all other Britons, took part in Great Britain's general election. Ballots cast in the election were sealed for three weeks to allow time for military service votes to be collected from all parts of the world. It wasn't until July 26, 1945, when the general election ballots were opened and counted, that the world reacted with stunned disbelief over the news that Winston Churchill had been swept from office in a landslide Labour Party vote. Clement Attlee, an unimpressive, dull speaker who was known for his plodding criticism of Conservative Party foreign affairs, was labeled by political enemy firebrands as "a snipe pretending to be an eagle." Yet the lackluster Attlee would be named Prime Minister, replacing the living legend Churchill, who had successfully brought vigor and unswerving resolve in a rallying call to his beleaguered countrymen. His oratory, his presence, his charisma, all of this was the acknowledged linchpin that played such a key role in winning the war.

Churchill had lost. Germany had lost. Riddle Field was lost as a throbbing, vital part of Clewiston's social and commercial fabric, even though that loss was celebrated throughout the streets, in the war-weary homes of the people of Clewiston, and throughout the rest of the nation. Lost also were those 201 cadets in Courses 25 and 26, who fretted over the "effing fickle finger of fate" that had denied them "a chance at the Huns," as Peter Spenser, an early Riddle cadet, had told his RAF evaluation board. The cadets' hope for combat gave way to reality late in July, 1945, when John Bennett of the British Ministry of Labour met with them to outline the bureaucratic

John Paul Riddle, an American civil
aviation pioneer, operated Carlstrom Field
and oversaw construction of the British
Flying Training School (BFTS) at
Clewiston, in Everglades country, in 1941.
His remains are buried near those of 23
RAF cadets at Oak Ridge Cemetery in
Arcadia, Florida. (Photo by Charles C.
Ebbets)

Aerial view of Riddle Field (north is at the top of the picture). Flight facilities are in
the lower center of the photo, and cadet housing in the upper left. (Photo courtesy
of Clewiston Museum)

plan of helping them face reentry into a peacetime economy (the eventual surrender of Japan had to be anticipated).

Bennett, the very model of decent mid-level government, was introduced by Wing Commander Charles W. Lindsay, who had recently returned from duty in Nassau to take command again of 5BFTS. Speaking to the restless, uneasy cadets, Lindsay opened by saying, "I know how you feel and perhaps you are due an apology for our ending the war with Germany in such a *short* time. I'm sorry, chaps, it's the best we could do." Laughter eased the tension and the highly admired Lindsay turned the meeting over to Bennett, who spoke on "Your Release and Resettlement." Bennett broke the ice quite nicely by telling a little joke: An economist pointed out how the war could be paid for if every adult smoked two packets of cigarettes and drank half a bottle of whisky daily. The taxes would keep Britain solvent, so why worry whether she was sober? Well-organized to a fault, Bennett spoke for perhaps one hour on the opportunities and options open to former RAF personnel. He answered a few questions, offered the services of his agency, and the meeting closed. When the cadets straggled out of the meeting room, an RAF Flight Lieutenant spoke to a few of the boys that he knew. "My, my," he said, "now you are redundant. What in the world will we do with you?"

With the end of the British flying training in the United States, there were enthusiastic plans for future development of Riddle Field. A civilian flying school, perhaps, or a municipal airport. Some saw it as a natural location for a manufacturing site, now that Uncle Sam was out of war production efforts and private business was gearing up to turn out consumer goods. Considerable interest in maintaining the splendid facility with its many amenities failed to produce the necessary financing to make Riddle a continuing viable entity. The old field became state property two years after the end of the war when the federal government turned it over to Florida. It soon became apparent that it would be a "tough sale" even to give the land away to a private developer. Gone were the days of the early 1940s when pressing wartime needs pulled government purse strings wide open with a cost-be-damned attitude. There was no cost ceiling, no budget, when it came to survival. But that was then and this is now, reasoned the number crunchers of the late forties and the entire fifties. The war was over, thank you, and welcome to reality.

Harvard AT-6 trainers from Riddle Field fly in formation. (Photo by Charles C. Ebbets)

Mr. and Mrs. Ira Nesmith (left and front center) at "RAF Headquarters Palm Beach," the home away from home for many British cadets. (Photo courtesy of Clewiston Museum)

RAF cadets willingly picked up the odd American ways . . . and happily introduced such quintessentially British pastimes as cricket to their U.S. comrades.
(Photo by Charles C. Ebbets)

The base canteen at Riddle Field was a popular spot with cadets. Accustomed to
severe rationing in Great Britain, the young RAF airmen were delighted with the
abundance and variety of food in the United States. (Photo by Charles C. Ebbets)

High hopes were shattered time and again as attempts were made to convert the monument to Anglo-American alliance into a viable peacetime endeavor. The city, county, and state listened to prospective developers who had glowing plans, but dim financial backing. After listening to a record-long presentation, one county commissioner said, "Hearing those pie-in-the-sky proposals is like listening to some guy with a bad haircut and a cheap suit trying to sell you on a million-dollar deal."

Many former cadets who had trained at Riddle have returned over the past years on sentimental journeys, always making it a point to visit the Clewiston Museum on South Commercio. George Cordes, the knowledgeable museum director, proudly points out the museum's RAF corner, where photographs, flying uniforms, and various other Royal Air Force mementos are displayed. Ray Searle, Course 18, has even given his house in Surrey, England, a name to evoke memories of those flying days above the dense canebrake, orange groves, and 'gator country: "Clewiston," it says on the front door. Michael B. Carroll, Course 11, visited the field in the 1950s and was quoted in a *Miami Herald* feature story. Carroll, then a business executive in Toronto, Canada (British-born), had been a Halifax bomber pilot. He was shot down over Stuttgart, Germany, on his second "op" and spent more than a year in a German prison camp. Carroll found that some of the old buildings were still standing, but all save one of the hangars had been razed. Guppies had been put in the algae-covered swimming pool to gobble up the mosquitoes. The once-excellent tennis courts, where tennis star Don Budge had coached the cadets, were deep in choking weeds. Old rusting aircraft components posed a threat to anyone foolish enough to take an evening stroll, and grimy mildew slashes violated the control tower, which had been a milky-white landmark for cadets coming home to base after a cross-country flight.

A few days after ex-Pilot Carroll returned to Toronto, a bulldozer showed up to take down some remaining structures. With the blistering June day hitting 90 degrees in the shade, a bare-chested operating engineer slammed the rugged equipment into a steel-rod-reinforced building. Again and again he rammed the building with little damage. "Son of a bitch," he yelled in frustration to a truck driver. "They say this field took a couple of months to go up. I sure as hell don't believe it. It just isn't possible." "It's one of those you-had-to-be-here things," the truck driver said. "Yeah, it really happened, it really did."

Yes, it did happen, and Riddle Field had a "good run" from September, 1941 until August, 1945. But with the glory days ended, Riddle Field was destined to become a blip in the history books, rolling over to saw grass, snakes, chunks of twisted metal, and fire ant hills. It is fair to say that the thousands of people who lived in the Clewiston area in those days will remember always the roaring engines, sunlit wings, and the young men who owned the skies.

Two years after Riddle Field graduated its last RAF cadets in August, 1945, lowered the Union Jack, and shut down flying operations forever, two devastating hurricanes slammed into south Florida. Both urban areas and farmland were flooded. Crops were hopelessly lost. Total damage exceeded $66 million. These two "big blows" of September and October, 1947 fueled a controversy likely to last well into the next century. From an ecological perspective, the Everglades were seriously threatened, with little hope for a partial, let alone complete, recovery. Primary blame was placed on lack of flood control.

Despite the slide toward total destruction, there has been a note of hope for a different type of flight activity at the old field, known today as Airglades Airport. Skydiving has captured the interest of many people searching for that quick rush of adrenaline when jumping out of an airplane at 8,000 feet. The airport shows hints of an even brighter future. The county-owned Airglades Airport has undergone an upgrading program. Each of the two runways has been lengthened to 5,900 feet, and two new T-hangars allow storage of ten general aircraft. The giant U.S. Sugar Corporation constructed a hangar at Airglades, following a decision of Clewiston city officials to close the Clewiston Municipal Airport.

By the mid-1990s, finger-pointing over ecological problems in the Everglades was divided between "self-interested" sugar growers and "ecological extremists." Environmentalists pointed to the generally accepted theory that the Everglades were threatened because its very life required a constant flow of clean, fresh water. They blamed the phosphorous runoff from sugar cane fields for fouling this flow. Sugar interests countered that many other factors contributed to the pollution—an exploding population and sprawling farms and ranches.

No single person other than Marjory Stoneman Douglas has

brought the Everglades into such prominence. Her 1947 book *The Everglades: River of Grass* became a classic. In 1987, the ninety-eight-year-old author who had sounded the alarm over the half-century deterioration of the Everglades called herself optimistic about the survival of this natural wonder. She wrote in a revised Pineapple Press edition of her book, with Randy Lee Loftis: "People often describe natural places—the mountains, waters, deserts, swamps and forests—as fragile. They mean to argue for their preservation. But they do them a disservice. Natural places . . . are not fragile. They are, in the main, tough as an old tire . . . the capacity of the earth for compensation and forgiveness has kept the planet alive. The Everglades is a case in point."

Profiles of Riddle Cadets

William I. (Bill) Davies, D.F.C.

Toronto, Ontario, Canada

In 1942 Bill Davies completed a sixteen-week Royal Air Force Initial Training Wing Course and was posted to elementary flight training school, flying Tiger Moths. With fewer than nine hours' instruction, he soloed on Tigers and was classified PNB (Pilot, Navigator, Bomb Aimer) at Heaton Park, a large transient depot in Manchester, England. Some years later, Davies went back to Heaton Park on a visit. Not a scrap of evidence remained even to suggest that the huge military camp had ever existed. Wrecking balls had knocked down barracks, recreational facilities, office buildings, and all the other structures that were spawned in Britain's effort to build the war machine that would spell survival. Davies learned that the local city planning authorities had rushed the demolition effort into action because it was feared that the old camp would remain in perpetuity. Heaton Park, with 650 acres, is one of the largest city parks in Europe.

Just before Christmas of 1942, we PNBs were sent by train to a destination unknown to us. It turned out to be Gurock, a small port near Glasgow on the river Clyde. We knew, of course, we were to board a ship and sail to either South Africa or Canada as part of the highly successful Empire Air Crew Training Scheme. South Africa or Canada? The general feeling was that we would head south rather than west, although there was no rationale for that feeling. After waiting day after day, we were finally transported by tender in the dark evening hours to a large mist-shrouded vessel. As we grew near the huge ship, someone shouted, "Holy Christ, look at that!" "That" turned out to be the *Queen Elizabeth.* I could make out her name as my eyes cut through the misty curtain.

Once underway, we zigzagged at a fast pace—twenty-eight to thirty knots—across the dark Atlantic. We were totally blacked out as we cut through an ocean that was a favorite hunting ground for the feared German submarine pack. German subs were sinking merchant vessels a few miles off Florida's east coast. The "rumor factory" was well underway after a day or so at sea. Someone had fallen overboard! Totally false, but the rumor had a life of its own and couldn't be stopped. The "someone" took on several identities. "I heard he was from Liverpool" . . . "he told a sailor that his mum had just died" . . . "what I heard was that he didn't want to fly." In the unfortunate event that anyone had toppled into the drink, there would have been no attempt at rescue. The troopship was committed to holding course and speed. Any foolish deviation could result in having the ship become a juicy torpedo target for the hungry U-boats. There could be catastrophic results in any change of plan.

I shared a cabin with two other flight trainees, one of whom was Pat Jackson, a graduate of a police college near London. Pat had managed to sneak himself a tiny role in the Noel Coward movie *In Which We Serve.* Jackson's part called for him to be clinging to a life raft alongside several other torpedo victims, all completely covered in oil. The movie shows German JU-88s diving down to machine gun the helpless British survivors, as Jackson says, "Here comes the bastard back again!" In the two or three times I have seen the movie, I have never been able to identify Jackson. He had told us several times that he would love to bail out of an aircraft.

After three or four days we disembarked at Halifax, Nova Scotia, boarded a huge Canadian train, and headed for Moncton, New Brunswick. We made a stop at Truro, where a group of charming matrons gave us baskets of delicious apples.

In Moncton we were told we would be doing our pilot training in the United States. We were given a list of the various British flying training schools and told to select our choice of locations. None of · my group had any strong feelings, so I volunteered that we should go to Florida, to Number 5BFTS at a place called Clewiston. We looked over the map to discover that it was at the southern point of Lake Okeechobee. Soon we were on our way south in a clean super-train, steaming along through Maine. We stopped in Boston and had the entire afternoon and evening to do as we wished. A welcoming committee of Bostonians was very generous to us, and we were soon guided around, admiring the beautiful buildings. We dined on superb food and had the best coffee I ever tasted. Back to the train, and another fast ride found us pulling into New York's Grand Central Station. We received orders to be back that evening at a certain platform and specific time. We saw the famed ship *Normandie,* which had burned while docked, raising speculation of arson. We were anxious to find Jack Dempsey's Bar on Broadway for a glass of beer. But Dempsey didn't show and we went back to Grand Central ready to board for the next leg of our journey. One of our cadets, Freddie Knight from Northampton, strapped on his accordion and we belted out the lyrics of "I'll Be Seeing You" and other popular songs of the day. Dozens of Americans joined in for a real sing-along. As we prepared to board our train, the crowd of Americans threw half-dollars, dollar bills, chocolate bars, and packages of cigarettes: Camel, Chesterfield, Old Gold, Lucky Strike, Phillip Morris, all new to us. We had a mass of loot!

We finally found our way to the train. This train was filthy and obviously hadn't been washed in weeks. The grimy windows were double-paned and jammed shut. Written on one of the panes was, "Joe for King" and "Tondelayo for Queen." We assumed that "Joe" was Joe Stalin and "Tondelayo" was Hedy Lamarr and that the scribbling in the window grime was done by a previous RAF Course many weeks ago.

Something very peculiar happened as our train headed south. Several cadets started to develop red spots all over their bodies and were taken from the train to local hospitals at every major stop along the way. What a mystery: one moment they were with us and the next they were gone. We never saw them again. Soon after we arrived at Riddle we learned that one of our group who had been taken off the train on our way south was diagnosed as having a very serious virus and was sent back to England. Another cadet developed severe melancholia and also was shipped back home.

It was a quiet, warm day when we got off the train in Sebring,

Florida, and entered a lovely small hotel where lunch had been prepared for us. There were enormous containers of fresh-squeezed orange juice, of which we consumed quart after quart. We crammed as much food as we could into our half-starved bodies and were staggered by the plenty. We had been living since September, 1939 in a heavily rationed country, Great Britain, and were amazed at the abundance of food in the States. Before we left the hotel, I read a sign on the notice board. It read: "OUR WAR EFFORT—ARE WE GOING ALL OUT?"

We were greeted at Riddle Field while a Royal Air Force standard flapped in a January breeze (85° F). The commanding officer, RAF Wing Commander George Greaves, welcomed us with a harsh speech on how to behave in our new surroundings. "Remember this," he said, "when in Rome do as the Romans do." We knew what he meant and were frankly dismayed by the unbelievable treatment meted out to the blacks by the whites. It was said of Wing Commander Greaves that he was a superb pilot and a scratch golfer who could also play a magnificent game of tennis. At one time he ran the 100-yard dash in ten seconds. He had all of that and more: he was universally disliked, the RAF's answer to Captain Bligh (*Mutiny on the Bounty*). One cadet put it this way: "He is the sort of Englishman who makes the rest of the world loathe all Englishmen."

Our group continued enjoying in our spare time the beautiful pool and I kept fit by doing what I loved: running middle distances. Finally, I clocked four minutes and fifty seconds for the mile. Later, in 1946, I did the mile in four minutes and thirty-six seconds while stationed in Malaya.

The extreme heat and punishing humidity made lectures a study in torture. During one lecture I woke up to discover that all the others in the class were sound asleep. I was the only cadet listening to the drone of the navigation instructor, who always called the computer "the confuser."

We flew Stearmans (PT-17s) that were somewhat more powerful than the Tiger Moths and I soloed very comfortably in about eight hours. Soon I was performing solo aerobatics, bringing forth a compliment from civilian instructor C. S. Presbrey. "Davies," he said, "you're going to be real good." My navigation and meteorology were quite good, and I was receiving marks in the high 70s and 80s. Word quickly spread one mid-morning that a Stearman had crashed and an unknown student pilot had parachuted to safety. "Pat Jackson!" I thought instantly, recalling that his fondest wish was to bail out of a plane. And, yes indeed, it was Pat Jackson, who gave a plausible story about having no choice other than bailing out. Eventually, he washed out and was sent to Canada for reassignment.

Tom Dixon, from Middlesborough, Yorkshire, was an interesting red-headed ex-policeman with an obsession about being commissioned when he received his pilot wings. He simply could not imagine graduating with sergeant stripes. Getting the commission seemed far more important than graduating as a pilot. Dixon was a member of our group who lived in the same quarters. Speaking with a very strong Yorkshire accent, he would invariably preface a remark by saying, "When I get my commission . . ." When Dixon wasn't present, someone in the group would impersonate his familiar chant and really take the mickey out of him. Tom Dixon, a good student and pilot, was completely humorless.

We could go to the mess hall at virtually any hour and order pretty much anything we wanted. I had made friends with a young black employee at the mess hall who was particularly helpful and couldn't do enough for us. It was obvious that this young man was extremely poor and a few of us used to tip him. I gave him the odd shirt and pair of slacks, for which he appeared extremely grateful.

Even though the facilities at Riddle were much more luxurious than we ever expected, we counted the days until the quarantine period ended. When that day arrived, several of us went into Clewiston and were entertained by the very generous local people. After a few days of being able to come and go pretty much as we wanted, three of us obtained long weekend passes and decided to visit West Palm Beach. Our plan called for us to hitchhike along the Tamiami Trail. We hadn't waited for more than a few seconds when a huge car stopped and a balding man in his late fifties told us to "come along." My two friends sat in the back and I got up front and chatted with the driver. The friendly driver asked where we would be staying in West Palm. When I told him that we would look for an inexpensive place for the three of us, he surprised us by saying, "Oh, now don't worry about that. I know a very nice hotel, and it won't cost you a dime!" We told him we couldn't accept the offer, but he was very insistent and, once again, we were amazed at the generosity of Americans. We motored up to the entrance of a swank hotel and were ushered to the front desk while our balding host whispered a few words with the desk clerk. We signed in and thanked the kind and generous stranger with the large car who shrugged off our fervent words of appreciation.

"No need for that, lads, just have a good time and it won't cost you a thing." What a marvelous country!

We spent a quiet and cautious weekend. I remember sitting at a table at the edge of the dance floor, watching the couples glide by. One bespectacled U.S. Army man was dancing with his date. It struck me as strange that the song lyrics included the phrase "back in nineteen forty-three" because that's what it was right then and there—1943.

Still basking in our good fortune in being helped by the kindly gentleman, we packed our grips and went down to the deserted lobby to check out. We were presented our bill by a hotel official behind the front desk. I asked about the "arrangements" that were supposed to have been made by the person who took us to the hotel. We grew up just a little when we learned that no arrangements had been made, and it took every penny we could scrape together to pay the bill. I have often wondered what motivated our balding "friend" to conduct such a charade. Egotism perhaps? I simply don't have the answer.

We had barely reached the highway that would take us back to Clewiston when up came a police car driven by a huge policeman armed to the teeth. (We found it fascinating to see those massive men with big guns in holsters with real cartridges around the middle). "Where you boys headin'?" he asked. When we replied "Clewiston, sir." he told us to "sit tight" for a minute or two. We waited until he spotted a large car and waved it over to the side of the road. The big officer with the big gun gently told the driver, "Now you be sure that you set these here boys down safely in Clewiston, you hear now." "Sure do, sir," replied the driver as we three gave our profuse thanks to the big cop. It seemed that we were home in no time!

Now that we could mingle with the Clewiston townspeople, we enjoyed their warmth, generosity, and hospitality. We were made to feel truly at home, and we enjoyed the contrasts of a fairly sleepy small Florida town with the totally different lifestyle we had left behind. In my mind's eye, I could see our green and pleasant land, England, pleasant in spite of the bombing.

Some of the aviation high jinks concocted by the cadets have become legends in the telling and retelling of stories that become more interesting as time goes on. One such incident involved two student pilots who were a course or two ahead of us and already flying Harvards (AT-6s). As students progressed they were to fly cross-country navigation exercises, one student flying as pilot and the other in the role of navigator. The assignment could well be a course consisting of four different legs, ending up back at base within possibly two or three hours. Needless to say, discipline in the air was very forcibly brought home to all students. Falling prey to the lure of low flying, always an incredibly exciting pastime, was grounds for expulsion. Tempting though it was, flying at treetop level or lower was extremely dangerous, not only to students and the aircraft, but also to innocent people on the ground. These two cadets set off on their navigation cross-country exercise. At some point over a very sparsely populated area they decided to do a little low flying. They took their Harvard right down to the deck, as the phrase had it in those days. They were zipping along so low that they bent the tips of the propeller and somehow managed to scrape by a tree and get a piece of branch stuck in the wing—big problems. When they recovered normal flight they knew their low-flying escapade would be found out the minute they returned to base. They would be instantly washed out and sent packing, with all of their studies and flying time gone for nought. After a long talk about their dilemma, they decided there was only one "honorable" thing to do: return to base and claim that their undercarriage wouldn't come down. They hoped to try a pancake landing, thereby disguising their propeller damage, and try to get away with it. Everything worked like a charm. Senior instructors in the tower talked to them on the radio, telling them to try this and that in an effort to get the wheels down.

"We've tried everything," radioed the pilot of the daring duo. "We just can't get those wheels down and we're really sorry and all that." Finally they were ordered to go ahead and try for a wheels-up landing. The landing was a brilliant success, damaging the prop (once more) and landing far down the runway. The two cadets scrambled out of the plane and surreptitiously removed the piece of branch from the wing before any motor transport reached them. It was a masterful cover-up. The pilot was congratulated and his log book was marked with a green [outstanding] commendation. Modest to a fault, the young pilot just gave a little smile at the many "well done's" that came his way. "It was really nothing," he said, "really nothing at all."

My instructor kept telling me how pleased he was with my flying progress. I believed Instructor C. E. Presbrey when he told me that I was going to be "real good." Those words gave me great confidence in my own abilities to handle the aircraft.

We spent more time socializing with the fine, friendly people of Clewiston, who opened their homes and hearts to us as we opened our hearts to them. Despite all our appreciation and respect for the people who befriended us, however, we could not come to terms with the harsh treatment meted out to the utterly subservient blacks. This truly was a paradox.

One day when I was alone I took a book along to the mess hall. I had decided to catch up on my reading while I ate lunch. The young black mess hall worker with the friendly manner and flashing smile hailed me. "Mister Davies, sir, I've joined the army and want to say goodbye." I instinctively got to my feet, shook his hand, congratulated him, and wished him the best of luck. I believe I gave him a few dollars without anyone seeing the gesture. When he left the mess hall, I sat down and resumed eating and reading. Then came a tap on my shoulder and I was advised that the adjutant wanted me in his office immediately. I couldn't imagine why I was wanted by the adjutant, a very good and soft-hearted man who tended to "mother" cadets. When I stepped into his office, I could see the pain on his face when I announced my name. To my complete amazement, two non-commissioned officers took places on either side of me and one shouted, "Left, right, left, right" and marched me directly before Wing Commander Greaves. His face contorted in fury, Greaves commenced to give me the biggest dressing down of my life.

I clearly recall seeing sprays of spittle from his mouth hitting my uniform as he shouted and raved at me. The odd thing was that I had absolutely no idea of what he was driving at. The thought that he was over the edge crossed my mind until it dawned on me that I had committed an unpardonable sin. Sure, the first words he spoke to our group when we arrived at Clewiston were, "When in Rome do as the Romans do." I had treated the young black man as an equal, something not tolerated in a white southern home.

Within a few days of this incident I was given an unscheduled flying test with an English RAF F/Lt., J. L. Crossley, and although I thought I had done quite well, I was told I failed and was washed out forthwith. I felt it was the end of my world. I sought out my extremely kind and understanding civilian flight instructor, Mr. Presbrey, who sadly told me there was nothing he could do. There was no doubt between us that we both understood exactly what had happened and why it happened. The adjutant urged me to go to Fort Lauderdale or West Palm Beach on a 48-hour pass and advanced me a little pay. I did go and spent a perfectly horrible weekend during the absolute lowest point in my young life. I was washed out and would never be a pilot. How could I ever amount to anything after this humiliation, which came about by merely being decent to a fellow human being? As it turned out it was the best thing that could have happened to me. To have been washed out turned into a blessing.

When I returned to Riddle Field, the adjutant handed over a railway warrant and money, with orders to report to Trenton Royal Canadian Air Force Base, Ontario. I was to be joined on the journey by Cadet R. J. Mackie, who also had washed out after many hours on Harvards in advanced training. We were told by the adjutant to take our time on the trip.

"Enjoy yourselves along the way," he said, "there's no deadline to reach Trenton." But taking one's time was not in Mackie's dictionary. He wanted to get to Trenton as soon as possible, thinking he would have a chance to obtain another flying test and continue his pilot training. (It never worked out that way and he wound up in some category of aircrew). We did manage to visit Savannah, Georgia, and Charleston, South Carolina, two cities that were utterly charming, particularly Charleston.

We arrived in Washington, D.C., which was completely swarming with people, making hotel space impossible to find on short notice. We came in the morning of a glorious day and saw all the usual sights and stood outside the iron railings on Pennsylvania Avenue staring at the White House, wondering what was going on inside. Accepting the fact that we would not be able to bed down that night, we wandered around not knowing exactly where we were heading. It dawned on us that we had not seen a white face for quite some time. There were many black people on the streets, and we were invited into an all-black restaurant, where we had a great southern fried chicken dinner. The owners would not hear of us paying. Everyone seemed delighted to be able to help us in any way and made it clear that they were intrigued with our accents.

After a 24-hour stop in Buffalo we were on our way again, finally reaching Trenton, our Canadian destination. We were assigned to other Courses that resulted in my receiving an "O" brevet (navigator half-wing) and three sergeant stripes at #3 Air Navigation School, Mount Hope, Ontario. We were shipped off very quickly to Moncton, New Brunswick, for a brief stay and were then taken to New York, where we boarded the *Mauritania*, bound for Liverpool.

After reaching England, it wasn't long before we were finished with the advanced training unit and on to the operational training unit. The Royal Air Force had a peculiar way of putting together the aircrews. They would funnel into a large hangar scores and scores of pilots, navigators, observers/bomb aimers, wireless operators, and air gunners (one for each crew). We were told, "Now sort yourself out into crews." Because we were assigned to clapped-out (worn out) and poorly conditioned twin-engine Wellington bombers, we would need a second air gunner and flight engineer when flying four-engine Lancasters or Halifaxes. I was invited to fly as navigator with an officer named Don Wallace, who had a large ginger-red mustache. Wallace was reported to have had a great many hours of flying time in South Africa, where he had been posted for a few years.

With the exception of Mike, a Canadian flying officer air gunner, we had an all-English crew. We later picked up two more Englishmen: Jim, a rear gunner, and Eddie, a flight engineer. We were to fly a four-engine Halifax. Don Wallace (Wally) had no difficulty converting over from the two-engine Wellington. After we flew twenty-eight hours over a two-week period, we moved on to Lancaster finishing school. The Lancaster had a slightly higher performance rating than the Halifax, but was somewhat more difficult to escape from in the event of emergencies.

We completed day- and night-training activities and were ready to operate against German targets by being posted to the 300 Polish Squadron at Faldingworth and briefed in Polish! Koenigsberg was to be the target, and it was expected that our crew would fly because one of the Polish crews had not yet returned from leave. Shortly before takeoff, the crew showed up and the CO ordered the experienced Polish crew back on the operation and we were stood down [taken off the operation]. Five aircraft were lost on that 29 August 1944 raid, including the crew that arrived late from leave and took our place.

Without explanation, we were promptly posted to 103 Squadron in Lincolnshire and were soon on operations. Our first and very lively target was the I. G. Farber chemical complex and we carried a 4,000 pound bomb called a "cookie," together with about thirteen small containers of incendiary bombs. Very quickly we did a second night operation, dropping twenty 500-pound bombs on the runways of a German night fighter station in Holland. The following day, carrying the same load, we bombed sites near the Holland coast.

Rumors spread from one crew to another that our squadron was to be visited by Group Captain Hamish Mahaddie, known as "Hamish the horse thief." His instructions were to go all around the Bomber Command and do whatever it took to persuade the top crews to join the Pathfinder Force (hence the term "horse thief").

Mahaddie took his marching orders from a tough-minded young Australian, Air Vice-Marshal Donald C. T. Bennett, chief of the Pathfinders. The Pathfinders came into being (and not without opposition) as simply a target-finding group that would go in ahead of a wave of bombers. They would locate the target and mark it with flares and incendiaries. The resulting fires would let the main force know where to drop its bombs. The assignment would be a difficult one, with an assumed high mortality rate, and the manning of the force would be determined by "creaming off" the best crews of the groups. When the idea of a select target-finding group was first proposed, Air Marshal Arthur T. (Bomber) Harris was loud and clear in voicing his objections. He was strongly opposed to creating an elite group with special insignia, feeling that it would impact on the morale of regular bomber aircrews. Harris, chief of the Bomber Command, finally accepted what would be known as the Pathfinder Force and suggested that Pathfinder crews would wear a brass eagle on the left-hand top pocket flap.

All RAF aircrews were volunteers and had to complete a tour of thirty bomber operations against the enemy. Many never reached that magic number. Generally speaking, the Pathfinders had a higher than average "chop rate" [killed or shot down] than regular aircrews. Once you were chosen to join the Pathfinders, you were made to sign a document agreeing to fly a minimum of forty-five operations. We were told that many crews had refused to go along with that stipulation because they felt that the target-marking aircrews had little chance of survival. But when our crew was told that we were the only one selected to be Pathfinders, we signed with gusto.

Almost overnight, we were posted to the Pathfinder navigation training unit at Warboys, near Huntingdon in the heart of Pathfinder country. At the conclusion of the extremely intense navigation training classes, I was made Navigator II. I would sit side-by-side with another navigator (no bomb aimer, of course) and I would drop all the target-indicator markers. Sometimes the drop would be made visually, but mainly "blind"—that is, by radar. Many other crews operated only visually, carrying a bomb aimer who dropped the target-marking flares from his position in the nose. Navigation was "king" in a Pathfinder group. That was made crystal clear to us in a meeting with Air Vice-Marshal Bennett, who stressed that accuracy was our key to success. He was blunt when he spelled out what he expected from us. "Remember this," he said, "I will have no compunction whatsoever in moving you back to the main force if you don't meet the Pathfinder standards." Rough-tongued and straight from the shoulder, Bennett was known as a flier's flier.

Going into early operations, we carried bombs and the bigwigs were not really concerned how accurately we dropped them; our main purpose was to find the target, be on time, and lend support to the master bomber, the first one over the marked target. On our seventh operation the target was Wilhelmshaven, on the North Sea. We carried six 2,000-pound bombs and once again acted as support. Thirty minutes before we reached the target, we were hit with heavy flak in the cockpit area, causing an electrical fire and vast clouds of acrid smoke. The bomb doors were opened and I was told to drop the bombs, but I refused. "We'll be there (at the target) in about twenty minutes," I said. The bickering continued (should we or shouldn't we) until Navigator I said, "Oh, I'll drop them," and reached over and let them go. Down went the bombs and the bomb doors were closed. When the bombs were released, a camera automatically went off on a timing device and shot a series of exposures where the bombs would strike. We were in plenty of trouble because we released those bombs before we should have. Here we were, going straight toward our target without any real problem after the flak incident. We reached the target and "agreed" that that was where the bombs had been released, rather than many minutes before. "We'll cook the books," Leo said. "If we don't we're going to be in the manure." I readily agreed to the plan and adjusted my log to fit his.

After we returned to the base and were debriefed, we handed in our logs. Then our crew set up a meeting without the presence of Wally, our pilot, who had given the orders to drop the bombs well in advance of the target. Mike, the Canadian flying officer gunner, led the attack on Wally. Several other crew members agreed with Mike and said they wouldn't fly with Wally. "Wally is strictly yellow," Mike said. "If we fly with him, we'll wind up dead." That seemed to be the tone of the entire group at that time.

Following a late breakfast, Leo (Navigator I) and I ambled into the navigation section about 10:00 A.M. There were two squadron leader bosses present named Dean and Blackadder. Dean, the senior of the two, looked around the room watching the men at ease who were playing cards or just talking. Then he crooked his finger at Leo and me, indicating we were to go into the office. We heard a mumble of remarks from other navigators, including "Oh, are you ever in the shit now."

Dean, about thirty-five, was an astute officer who had heard all the stories and excuses in the book and was not about to be taken in with flights of fancy. His counterpart, Blackadder, maybe a year or so older, also was impervious to tall tales.

"All right," Dean said, "have you anything to tell us?"

Leo and I feigned surprise, mumbling and stumbling over words of denial. But Dean, losing his patience, exploded and pounded the table: "Do you think we are children and don't know what is going on? We know what the hell happened. Plain and simple, you cooked your fucking logs and now what do you plan on doing about it?"

Leo opened up right away, telling the full details and more than was necessary, while I followed suit. Our attempt at cover-up was over and what we did, God help us, was in full view.

"So what do you plan to do?" asked Dean. We told him that we had another crew meeting, without our pilot, scheduled for 2:00 A.M. Dean stood up and dismissed us: "Very good, gentlemen, please let me know where you want this thing to go."

Our final decision was to support our pilot, Wally, all the way, with only Mike, the mid-upper gunner fighting our vote. But Mike was the odd man out. We told Dean and Blackadder that Wally was our man, and that was the end of it. Nothing more was said.

We flew on several more night operations and then were attacked by ME-110s on a trip to Düsseldorf, taking on several hits, but managing to evade both times. As more and more operations followed, we became a more skilled and disciplined crew. Wally, our pilot, developed into a very good captain, instilling confidence in the crew as we took on jobs that presented much more than ordinary challenges. Blind illuminators involved a technique for dropping large, long-burning magnesium flares, using radar when the target was obscured. Planes carrying visual bomb aimers would fly behind us at a lower altitude, and drop their markers on targets identified by the brilliant magnesium flares.

One wintry night early in February, 1945, we were briefed to fly to Stettin, Germany, on the Baltic. We expected to fly in our favorite aircraft, "T" for Tommy, but had to settle for "A" for Apple because Tommy had some sort of problem. En route to the target over Germany one engine began to overheat, and we closed it down and feathered. Our target was an oil refinery at Politz on the outskirts of Stettin. We were about 18,000 feet in bright moonlight on a bitter-cold night. I dropped blind illuminators and the visual bomb aimers marked very accurately. Reports pronounced the raid a huge success. The refinery was destroyed and produced no more oil during the war. After the drop we turned east and headed toward the Swedish seaport city of Malmö. I left my navigation compartment to get a better look at Malmö, for it was all lit up, the first illuminated city I had seen in quite some time. I clearly saw traffic lights in the busy streets change from red to green to amber. Flying over neutral Sweden was internationally banned, but we were routed over a corner of it. A Swedish artillery battery did send up some flak but deliberately (we thought) exploded the shells at 10,000 feet, well below us.

We changed to a more westerly course, flying sedately along with one of our four engines out of service. Suddenly Mike, the mid-upper gunner, screamed that a JU-88 was sitting on the starboard beam with his cockpit light on. "Watch out," he shouted, "there's another one coming up behind all blacked out." I had barely scrambled out of the navigation compartment when I clearly saw the JU-88 and pilot as we were raked with a long burst of cannon fire. Chunks of metal were ripped from our plane and one engine was knocked out. We were corkscrewing madly on two engines and our gunners were firing the .303s almost nonstop. Another long enemy cannon burst and a second engine was blown out. That engine provided the power for gunners to raise or lower their guns. With only one engine operating we made a frightening dive almost straight down to about 2,000 feet, when Wally did a tremendous stomach-dropping, shuddering pullout and leveled out in darkness and slashing rain. We concluded we had lost the JU-88s, but knew that the Lancaster (magnificent though it was) could not maintain altitude on one engine, particularly with all the damage it had suffered. We had about four hundred miles of water to cross and were flying at only about one hundred and fifty feet above the icy North Sea.

Wally told John, the wireless operator, to send out an SOS. "I simply cannot maintain height, and we shall have to prepare to ditch." John replied that an SOS at our height would be useless, as no one could receive it. Eddie, the flight engineer, suggested that we try to restart the closed-down feathered engine. Wally agreed and they went at it while we were hoping against hope that it would start. What a glorious sound as that Rolls-Royce Merlin engine roared into action! Had not the engine restarted, our chances of survival would have been slim. Even if we had made a successful ditch (highly unlikely, given the circumstances), we would have had an almost impossible task getting into the inflatable dinghy. And if all of that had fallen into place, we couldn't have lasted in the freezing sea with the biting wind and huge waves.

With two engines purring sweetly we interrupted the control tower at base while some beat-up Lancasters were requesting emergency landing permission because they had only three engines. "Listen to this," Wally told the tower, "we're on only two engines and request priority landing." We were given the first-down approval and were the first aircrew to be debriefed.

After we had given our full report and answered dozens of questions, I suggested to Mike that it was awfully warm indoors and asked if he would like to go outside for a cigarette. I lit up and promptly fainted dead away. The next thing I remember there were two pretty young WAAFs loosening my collar and saying, "Look, he's coming to now."

Three nights later, we were returning as blind sky markers from Chemnitz. On our long journey home under a brilliant moon, many friendly aircraft were visible. Then came an attack. Many of our planes were shot down, some exploding, as Mike shouted his warning: "Corkscrew port, GO!" Wally put the Lancaster into an extreme left corkscrew as I jotted down in my log the altitude, position, air speed, and course we were flying. I then got out of the navigator compartment to watch the attack. I looked to the rear and saw an enemy aircraft with tremendous closing speed coming at us. It was the first German jet we had seen, a twin-engine ME-262 equipped with tracer cannon. Its first long burst of pale blue tracers just missed our port tail fin as our two gunners hammered away at the jet. I watched with awe as our curved stream of fire struck the jet and the starboard wing fell away. The jet fell sharply to the left in its tumble to earth and there were no parachutes observed.

I made a firm descriptive note in the log, giving all pertinent details, and we claimed the jet as a kill! Unknown to us at the time, another Pathfinder aircraft had witnessed the entire incident and put full details in their log, confirming our claim. Several weeks later we were officially confirmed for the kill and our ground crew painted a small swastika on our plane just below the pilot's position. These were rough, costly times, with many of our close friends lost to flak and fighters. We had four squadron commanders in a three-month period—three were shot down and one crashed just before he was ready to land at base.

Our final fighter encounter was really weird. Coming back from a deep penetration into Germany, we were about fifty miles due south of a German night fighter base. John, the wireless operator, was manning a radar lookout beamed to the rear and said there were two enemy aircraft closing fast. I came out of the navigator compartment and saw two ME-109s performing curve-of-pursuit attacks on our Lancaster. Bomber Command procedure called for us not to open fire until we were fired upon because the German fighters had cannon power, while all we had were miserable little .303s. Also, it was possible they had not seen us, although in this case it was more likely that they had. The two ME-109s continued to follow us for about ten minutes without ever opening fire. They finally veered away, and we never came up with any sort of realistic explanation.

One gloriously sunny morning, we were returning from a German operation, enjoying a cup of coffee at about five hundred feet off the North Sea and cruising along under 10,000 feet, having removed our oxygen masks. The crew was rather exhausted and our normal alertness had lost its edge. Wally decided to fly over a "fishing boat" and waggle his wings. It was about 6:00 A.M. and I was standing behind the engineer, enjoying the beautiful morning view. Wham! The "fishing boat" opened up with a withering barrage of antiaircraft fire and scared the daylights out of us. It was a German E-boat, fairly well-armed, and its purpose was to play havoc with returning bombers. We flew over Berlin and were glad to get that over with. Even today I get a feeling of fright hearing the names Frankfurt, Wilhelmshaven, Düsseldorf, Essen, Karlsruhe, Cologne, Zeitz, Stettin, Leipzig, Chemnitz, Nuremberg, Hamburg, Hanover, and Kiel. I completed forty-three operations, plus a food-dropping trip to a field just outside The Hague. That trip was very rewarding because we knew that Dutch families were near starvation. I had been commissioned, decorated by the king, and could wear the coveted Pathfinder award under my medals. Following a glorious rest, the RAF sent me on a course in England and then I was shipped to Saigon, French Indochina, in December, 1945. Saigon was known as the Pearl of the East, a really lovely city. One day I was strolling near Rue Catinat, the main drag, when I came face to face with Tom Dixon, the Riddle Field cadet who was obsessed with getting a commission. He was in the uniform of a warrant officer and had pilot's wings but nothing in the way of medals. His first words were, "How did you get your fucking commission?" We sat at a little table outside a cafe, and he told me his tale of woe over coffee. He had graduated as a sergeant pilot, missing out on the commission he felt sure would be his on graduation. He and his group had been promised that they would be flying gliders, transporting troops over enemy lines, but that never came through. Most graduates in his Course did very little during the war. Dixon was completely disappointed with his lack of achievements in real combat, and I genuinely felt sorry for him. When we said good-bye, I said to myself, "Well there's a lesson

here. Dixon was on top of the world at Clewiston while I thought
my world had ended. And now look."

I took my leave in Hong Kong, specifically to attend the Japanese
war trials being held in Kowloon. Attending the trials every day, I
was mesmerized by the parade of witnesses who described in
graphic detail the unbelievable sadism and cruelty that took place
behind those "bamboo cages."

I was offered a position on the staff of Air Vice-Marshal J. D.
Breakey at Air Headquarters, Malaya, in Kuala Lumpur. This was a
real plum and gave me a roving commission in Southeast Asia, as
well as a promotion from Flight Lieutenant to Squadron Leader. But
not so fast—no sooner had the Indian tailor put on my new rank
than I was told that my position had been downgraded to Flight
Lieutenant. When I told my tailor to go back to the F/Lt. rank on my
uniforms, I thought he would burst into tears.

At my new post, I flew to Borneo, all over Malaya, and once again
visited Saigon and Hong Kong. At the request of Lord Killearn, the
food commissioner in Southeast Asia, I extended my commission for
six months. Finally it was over and I enjoyed a sail from Singapore to
Southampton via the Suez Canal. I was demobilized after arriving
home in England in January, 1947, and I emigrated to Ontario in
June, 1948.

I have never regretted the experience for one second.

Robert G. F. Lee, D.F.C.

Shalford, England

*Robert Lee was a graduate of Course 4, Riddle Field, winning his wings on
May 2, 1942, and named the best all-round cadet among the thirty-nine
graduates at Wings Parade. Posted to night fighters, Lee crash-landed in
France on August 14, 1944, when his aircraft's engine was knocked out by
German ground fire. Badly wounded and trapped in his plane, Lee was
found by American forces eight days after he had crashed. He had no food or
water during an ordeal that received wide attention following publication of
a syndicated column by Ernie Pyle, famous American war correspondent for
Scripps Howard. Pyle was with an armored unit during a lapse in fierce
fighting that had left many American infantry units with heavy casualties in
a section of northern France. Some soldiers walked away from the group to a
low hedge about thirty yards from the road. In Ernie Pyle's words:*

The hedge was low and we could see over. There were two British fighter planes. One lay right side up, the other lay on its back. We were just ready to go turn around and go back when I spied a lone soldier at the far side of the farm field. He was standing there looking across the field at us like an Indian in a picture. I waved and he waved back. We walked toward each other.

He turned out to be a second lieutenant, a graves registration officer for his armored division, and he was out scouring the field for dead Americans.

He was glad to see somebody, for it is a lonely job catering to the dead.

As we stood there talking in the lonely field, a soldier in coveralls with a rifle slung over his shoulder, ran up breathlessly, and almost shouted: "Hey, there's a man alive in one of those planes across the road. He's been trapped there for days!"

We stopped right in the middle of a sentence and began to run. We hopped the hedgerow, and ducked under the wing of the up-side-down plane. And there, in the next hour, came the climax to one of the really great demonstrations of courage in this war.

We ran to the wrecked British plane, lying there upside down, and dropped on our hands and knees and peeked through a tiny hole in the side. A man lay on his back in the small space of the upside-down cockpit. His feet disappeared somewhere in the jumble of dials and rubber pedals above him. His shirt was open and his chest was bare to the waist. He was smoking a cigarette.

He turned his eyes toward me when I peeked in and he said in a typical British manner of offhand friendliness, "Oh, hello."

"Are you all right," I asked, stupidly.

He answered, "Yes, quite. Now that you chaps are here."

I asked him how long he had been trapped in the wrecked plane. He said he didn't know for sure as he had got mixed up about the passage of time. But he did know the date of the month he was shot down. He told me the date. And I said out loud, "Good God!"

Wounded and trapped, he had been lying there for eight days!

His left leg was broken and punctured by an ack-ack burst. His back was terribly burned by raw gasoline that had spilled. The foot of his injured leg was pinned rigidly under the rudder bar.

His space was so small he couldn't squirm around to relieve his own weight from his paining back. He couldn't straighten out his legs, which were bent above him. He couldn't see out of his little prison. Not a bite to eat or a drop of water. All of this for eight days and nights.

Yet when we found him his physical condition was strong, and his mind was as calm and rational as though he were sitting in a London club. He was in agony, yet in his correct Oxford accent he even apologized for taking up our time to get him out.

The American soldiers of our rescue party cussed as they worked, cussed with open admiration for the British flier's greatness of heart which had kept him alive and sane through his lonely and hope-dimming ordeal.

One of the soldiers said, "God, but those Limeys have got guts."

It took us almost an hour to get him out. We don't know whether he will live or not, but he has a chance. During the hour we were ripping the plane open to make a hole, he talked to us. And here, in the best nutshell I can devise from the conversation of a brave man whom you didn't want to badger with trivial questions, is what happened:

He was an RAF flight lieutenant, piloting a night fighter. Over a certain area the Germans began letting him have it with ground fire.

The first hit knocked out his motor. He was too low to jump. So—foolishly, he said—he turned on his lights to try a crash landing. Then they really poured it on him. The second hit got him in the leg. And a third round cut right across the tips of his right-hand forefingers, clipping every one of them to the bone.

He left his wheels up, and the plane's belly hit the ground going uphill on a slight slope. We could see the groove it had dug for about fifty yards. Then it flopped, tail over nose, onto its back. The pilot was absolutely sealed into the upside-down cockpit.

"That's all I remembered for awhile," he told us. "When I came to, they were shelling all around me."

Thus began the eight days. He had crashed right between Americans and Germans in sort of a pastoral No-Man's Land.

For days afterwards the field in which he lay surged back and forth between German hands and ours.

His pasture was pocked with hundreds of shell craters, many of them only yards away. One was right at the edge of his wing. Both sides of the fuselage were speckled with hundreds of shrapnel holes.

He lay there, trapped in the midst of an inferno of explosions. The fields around him gradually became littered with the dead. At last American strength pushed the Germans back, and silence came. But no help. Because, you see, it was in that vacuum behind the battle, and only a few people were left.

The days passed. He thirsted terribly. He slept some. Part of the time he was unconscious; part of the time he undoubtedly was delirious. But he never gave up hope.

After we got him out, he said as he lay on the stretcher, "Is it possible that I've been out of this plane since I crashed?"

Everybody chuckled. The doctor who had arrived said, "Not the remotest possibility. You were sealed in there and it took men with tools half an hour even to make an opening. And your leg was broken and your foot was pinned there. No, you haven't been out."

"I didn't think it was possible," the pilot said, "and yet it seems in my mind that I was out once and back in again."

That little memory of delirium was the only word said by that remarkable man in the whole hour of his rescue that wasn't as dispassionate and matter-of-fact as though he had been sitting at the end of the day in front of his own fireplace.

Military censors did not permit Pyle to reveal the name of the RAF night fighter pilot until some days later, after proper authorities had been notified. Pyle then reported that the subject of his column was Flt. Lt. Robert Gordon Follis Lee. Ernie Pyle, known by American troops as a "GI soldier's best friend," was killed by a Japanese sniper's bullet on April 18, 1945, while covering action of the U.S. Army's 77th Division on the Pacific Island of Ie Shima. Flt. Lt. Robert G. F. Lee spent many months in hospital recovering from a shattered left leg and severe wounds. His father, Frank Lee, reported in April, 1945 his son's progress. The senior Lee wrote to Riddle Field officials, stating, "I am pleased to say that Robert has recovered from all his wounds apart from the injury to his left leg, and this will, I am afraid, still take some time before it is well and of use to him."

Alastair Michie

Wareham, England

Alastair Michie was a Course 3 Royal Air Force cadet at Riddle Field. Following graduation he was presented with a silver bracelet inscribed "A. Michie, best flying cadet Course 3, 5BFTS, Riddle McKay Aero College."

Born in France, Alastair Michie was educated in France and Scotland and studied architecture at Edinburgh University. In the early 1960s he directed his talent toward fine arts and sculpture and has attained world prominence in the art world. His work is in private collections in Europe, the United States, and South America, as well as in museums, universities, and major corporations. Michie was one of fifteen British painters chosen to exhibit in Warsaw's Museum of Art in 1967. In 1972 he was invited by the government of Brazil to give a series of major exhibitions.

My first impressions of Clewiston and 5BFTS were: How could normally pale young British men in a course just ahead of us get so tan in such a short period of time? . . . dispensers of ice-cold Florida orange juice, all you wanted at any time . . . the Commanding Officer's welcome . . . a very low-level climbing roll over our heads in a flashing silver Harvard (AT-6A) . . . a hurricane in the area and some aircraft are tethered by wire and rope while others are flown away from impending danger.

All the billets and the dining room had just been completed and the barren land around us was enhanced by the planting of young palm trees kept in place by stakes and cables. We were fed in the immaculate new dining hall by colored staff in crisp white cotton jackets. We were introduced to iced lemon tea, hot cakes, and maple syrup jostling on our plates with bacon and eggs, all of which delights I cling to even now.

We had wonderful weekends in Miami, courtesy of the always-generous drivers offering transportation. The town of Fort Myers played host to us one weekend and lavished upon us great hospitality—wine, food, and girls. Throughout our time in Florida, generosity knew no bounds. They looked upon us as young heroic figures from a country under siege. I acquired a snakeskin belt from a Seminole Indian and thereafter never flew without it, to a point that I became deeply superstitious concerning it. A friend and I were given a lift one Friday night and persuaded to stay for a couple of nights in our host's elegant beach house. Saturday morning we were plied with three powerful cocktails and I recall very little more except drifting very happily throughout the weekend, enveloped in a soothing alcoholic haze.

Our instructors were the best, so highly skilled and patient. Examples included John S. W. ("Slow Walker") Davis, R. F. Hosford, and J. F. Reahard.

Cadet Albert Sloman and I were chosen to give a display of aerobatics on graduation day [March 12, 1942]. I rehearsed for the event by taking some ridiculously dangerous chances, and only the meticulous teaching of Johnny Davis saved me from disaster.

In five months it rained only once during night flying. Then dense fog set in, causing one cadet, Robert B. Crosskey, to stall too high and suffer a fatal crash in his BT-13 on January 20, 1942.

On my return to England I was commissioned and learned to fly twin-engine aircraft at the operational training unit: Oxfords, Blenheims, Bisleys, Havocs, and Bostons (A-20a). I was then posted to 605 Squadron, flying night intruders during two operational tours, about eighty missions. We flew mostly Mosquito fighter bombers with four 20 mm cannons. Our tasks were quite varied—to patrol German night-flying bases; to harass, bomb, and destroy targets of opportunity; and to search the seas for mine-layers, all at low-level and mostly at night. I flew twice to Berlin at 20,000 feet, mixing with the bomber stream and attempting to intercept German night fighters. I also flew in low-level fighter support of Lancaster Bombers (617 Squadron dam busters) attempting to breach canals.

I was stationed in East Kent when the first V-1 rockets were directed toward London. We would patrol off the French coast at about seven thousand feet where we tried to intercept them, successfully at times. They were fast, of course, and flew at about five hundred feet so that our high-speed dive was required to catch them.

After the war I resumed my architectural studies at Edinburgh University, but became impatient with the long training still ahead and went to work as a freelance graphic artist and illustrator until the early 1960s, when I decided to pursue fine art as a career.

David Stewart

Hamilton, Ontario, Canada

David Stewart served in both the Royal Air Force and the Royal Canadian Air Force. He was born in Scotland and emigrated to Hamilton, Ontario, with his parents when he was four. He returned to Britain in 1938 just before the war, first to Scotland and then to England, where he joined the RAF when war broke out. When hostilities ceased, Stewart returned to England, determined to marry the love of his life and take her to Canada. His father-in-law-to-be had a different idea, suggesting that Stewart return to Canada, get a job, and then come back to England and get married. Stewart took this suggestion and went back to Canada in 1946, staying there less than a year. In 1947 he was back in England; he rejoined the RAF and served until 1952, a happily married man. After reading of his many pranks and prangs, one must conclude that his father-in-law had a level head on his shoulders.

I lived north of Plymouth when the blitz came and it made a tremendous impact on me—the searchlights, the bombing, the red skies. As Jerry flew above us, controlling the skies, I was angered and impatient. I had volunteered for the RAF and was restlessly awaiting my call-up.

One day I went into Plymouth to find my workplace had been bombed and about 200 feet of the building had been ripped apart and was lying in shambles. Soon after, the RAF called me up and I was given initial training in the intriguing Tiger Moth biplane. I did all right, managing to get both wings in the air at the same time. To show their appreciation, the RAF brass chose to reward me with a six-month vacation in Florida. Furthermore they elected to pamper me with an oceanic cruise on the *Queen Mary*. Nothing but the best, especially because we were permitted to play "Dodge the U-boats." Splendid blokes, the Royal Air Force.

A troop train from Canada brought us into Sebring, Heaven, also known as Florida. We arrived at the Sebring Hotel, just in time for breakfast. Perfect timing to enjoy unlimited orange juice, sausages, bacon, eggs, pancakes, real bread, and anything else we wanted. In England aircrew members were privileged and could get one egg a week.

And sunshine! After completing our free stamped postcards to England, we were on our way to Clewiston. Thank you, Mr. Higgins, Sebring Hotel. Thank you, Florida, and thank you, Embry-Riddle for supplying "air conditioning" in an open cockpit PT-17 at 5,000 feet. Good way to beat the heat. When not flying we enjoyed food, glorious food, and the trips into Clewiston, frequently accompanied by "Queenie," our faithful mascot dog. Queenie seemed to know the bus schedule to and from Clewiston. We also got weekend passes now and again. Most of the boys chose West Palm Beach but I opted for Sebring, the proven town of generosity. I was talked out of Sebring by a man who gave me a lift. "I'll tell you, son, you ought to go to Fort Meade, it's a great town for uniforms." I accepted the suggestion and have never been sorry. I made friendships there that have weathered the years and currently remain intact. Sitting on the ground with two maidens peeling oranges as fast as I could eat them, interrupted only with time out for swimming, I thought, "This is the way to spend a war!" Anita Oakes's father would drive me part way when it was time to go back to the field. He would let me out on the highway, and I would get a ride to Riddle Field, usually from the first car that came along.

Our instructors were wizard chaps [wonderful guys]. Most of them survived until we graduated. On completion of our Stearman course I flew one vertically upward, holding her until she had zero airspeed and suddenly flipped over. This was a prohibited maneuver due to the possibility of breaking one's neck. I wasn't aware that the engine was being partially starved for oil—after all, I felt that I owned the world and everything was great.

My next and final trip in a PT-17 involved the same aircraft and was a farewell courtesy ride with my instructor. I flew at 200 feet and tried to lose him, but I didn't. Then I put her down for some low flying (and I do mean low). Without even a warning sputter, my engine quit. I pulled back to climb and look for a landing spot, but due to lack of altitude could only land straight ahead. After a successful prang [crash], fixed undercarriage and no flip-over, I looked in the mirror and saw the serious look on my instructor's face. Then he saw me reflected in the mirror and smiled. Hooray! I was forgiven. Quickly jumping out of the cockpit, I ran around to the front of the aircraft, fully expecting that the propeller would be bent backwards. I thought I had hit a cow a few miles back. A miracle. The prop was straight and I was innocent. I felt the prop and learned that the engine had seized, apparently the result of my previous vertical flying escapade that robbed the engine of oil. My belated apologies to 5BFTS and Embry-Riddle. I had an exam scheduled for that afternoon and requested I be given due consideration because of my "state of shock" after almost getting killed by treacherous aircraft through no fault of my own. The exam was on Morse code and I got 100 percent of the questions right. (I should have saved my "state of shock" for a tougher exam!)

I graduated to the sacred half of the field, flying Harvards (AT-6s). We no longer were the junior course. It was our turn to be worshipped. Harvards were real airplanes with retractable undercarriages. We flew in three formations of three followed by another nine aircraft. I was number three (portside aircraft) of yellow section (portside section). I had trouble with my undercart [landing gear] and couldn't raise it like the other pilots. As we approached the field, I was flying full throttle and fortunately was just holding my position because of the drag caused by my "unretractable" undercarriage. My Florida friends attending the graduation ceremonies were able to identify my aircraft.

Honestly, I didn't do it intentionally—but I did draw attention intentionally when I "shot up" Fort Meade, diving on familiar houses and other landmarks until residents of the town turned automobile headlights on the baseball field in case I wanted to do an emergency landing. A nearby U.S. Army Air Corps field was alerted to the "aircraft in trouble" and promptly sent up a plane to assist me. I saw the fighter's lights, and put mine out. Then I did a 270-degree turn so he wouldn't see my exhaust and headed home. When I touched down at Riddle Field I was innocent once again and surprised that there was no commotion. The Acuffs lived on the edge of town and twelve-year-old Bill climbed on the roof of the garage and waved at me as I flew by. I dived at him, apparently too low for his liking, as he jumped twelve feet to the ground. Many years later I talked to Bill and he reiterated his fears at seeing me do a slow roll as the engine cut out, backfired and belched smoke from the exhaust. As I righted the aircraft, the engine came on again. I learned then and there to pull up higher when doing a horizontal roll near the ground.

I flew a long-distance navigation trip and stayed overnight at the destination. Returning to base the next day after an intense, concentrated flight I was finally in position to request landing instructions and did so, adding "Home sweet home." A female instructor with a lovely voice gave me instructions, concluding with a warm "Roger and welcome home." It seemed like she kissed me at 2,000 feet.

While on operations in Burma, I got leave and flew to Karachi in western Pakistan. I arranged with an American crew to drop me off at Miami, but unfortunately became gut-wrenchingly ill from some Indian food, and the crew had to leave without me. So much for my

Florida leave. What would have happened if I had become ill after I was dropped off at Miami? The American military police would have had a field day finding me 10,000 miles away from my squadron with no transcontinental leave authorization in my pocket.

But then the world was so small to me, and Florida so attractive. When I have the opportunity to talk to some of my fellow 5BFTS trainees, we always seem to have a meeting of minds when we recall the Clewiston days. Our instructors did an excellent job, considering what they had to work with. I'm equally sure these same instructors would have doubted their teaching abilities and stood open-mouthed had they witnessed my engine cutting out at 2,000 feet at night over the Everglades, a prohibited area. The engine came back in at 1,600 feet—just as well because I was set to bail out at 1,500 feet, scared though I was of 'gators. I behaved after that, telling myself that these instructors know what they are talking about.

After the war, I was flying a single-engine jet fighter, the Vampire. I led six aircraft in search of an army convoy, following a road that snaked its way up the hills on the Yorkshire moors. As the road went into the low-lying clouds, I turned toward the coast, leaving four aircraft orbiting Redcar. I proceeded with my number two in battle formation (100 yards on my starboard), around the coast onto the moors under a dangerously low ceiling. As we approached where the clouds touched the ground I said, "Red 2, when we go into the cloud put your nose down." We were flying at twenty feet and he had no time to reply. I went into the cloud, put my nose down and broke the cloud in the valley. I looked to my starboard and there was Red 2, just where I knew he would be. Incidentally, "Red 2" is now Air Chief Marshal, the second-highest rank in the Royal Air Force. When flying Spitfires in England I pranged one when my tail wheel broke off during a strong crosswind landing. After returning to Canada, the RCAF arranged for me to fly Mustangs, which had a much wider undercarriage than the Spitfire. In addition, the Canadian weather was much gentler.

I got smarter as I got older and kept my private flying license into the 1980s with almost 40 years of flying, four prangs, and getting shot up twice. What's that about getting shot up twice, you're wondering? Oh, that was during my Burma days, when our plane was hit by antiaircraft ground fire. We were dropping supplies to our troops and the Japanese were so close to our dropping zone that they most likely got some of our supplies. They would fire on us as their way of saying, "Thank you."

Peter Brannan
Toronto, Ontario, Canada

Peter Brannan was a Royal Air Force cadet in Course 25 at Riddle Field, and later placed in Course 26 because of days lost due to illness. After the war Brannan emigrated to Canada, became the editor of Canadian Aviation *magazine and traveled worldwide in search of aviation stories. He flew to the Distant Early Warning (DEW) line in the Arctic and flew in the Concorde, as well as taking off and landing on an aircraft carrier. Brannan has met many of the giants of aviation history.*

I look back on those years with gratitude to the United States Army Air Force for their efforts during World War II and the great reception we British cadets were given in Florida. Aviation has pretty much been the second love of my life. In fact I became involved in flying long before I met Anne, my wife-to-be. Briefly, I am an ex-newspaperman. I built and flew model planes from the age of twelve and, at fifteen, joined the Air Training Corps, a British pre-RAF service organization. In 1943, when I was eighteen, I joined the RAF and soloed in less than two months.

Between training courses in England we did odd jobs on bomber airdromes in East Anglia. I was one of the many in the pilot/air crew hopper waiting for action, and happily the losses were obviously fewer than anticipated. While others were on bombing missions over northern and southern Europe, I was driving a tractor, helping to load bombs into Lancasters in the wee small hours. After we watched the "Lancs" take off we would see the masses of Flying Fortresses (B-17s) fly over from their East Anglia bases. After a year of such frustrating chores, I finally gained freedom and was posted to Canada for training. In February, 1945 I was on the *Queen Mary*, bound for New York and then Canada. I was one of the fortunate few chosen to train in the States and after traveling several days by train, arrived at Riddle Field, Clewiston, Florida, sometime in March, 1945.

Overwhelming is the only way to describe our reception from the people of Clewiston and West Palm Beach. They loved us and we loved them. Unfortunately, I have never found any trace of the people who were so good to us, although I did exchange Christmas cards for many years with our main hostess, Mrs. Ira Nesmith.*

But Florida, oh Florida. We were taken to shows and to church, and they fed us (oh, did they!). Until we arrived in Florida we had not seen a banana for some years and had almost forgotten the taste and texture of chocolate (dear departed Cadbury) and ice cream. They feted us and even turned over their cars to some cadets. I could fly a plane but when it came to driving, forget it. There was once, however, when some of us got tied in with some crazy U.S. Navy guys and their girlfriends one weekend. They had rented a car and were dedicated to doing much more than driving. They "volunteered" me to drive inasmuch as I was the only nondrinker at the moment. It also didn't seem to matter that I was the only nondriver in the crowd. What an experience—going 60 mph or more down Highway A1A without knowing what I was doing—completely sober but completely inept at the wheel. The car had an automatic transmission, and for that I thank the inventor of that miracle of motordom.

I was making progress in Course 25 at Riddle, but then was in sick bay and given a week's leave. I was free on my own, and I could feel West Palm Beach beckoning. Another cadet and I packed up and headed for the exit gate at Riddle Field with our thumbs in the appropriate position. With a squeal of brakes the first car that approached picked us up (this was typical) and took us to West Palm Beach. Our beacon was the home of Mrs. Nesmith, who greeted us like long-lost sons and despatched us to the winter home of the "Countess Apponyi" on Everglades Avenue. At the proper address, we found that the countess was not in residence at the time, but we were to be guests of the house. Countess Apponyi's Filipino manservant catered to our needs—cooking, making our beds, and keeping the fridge

---

*Mrs. Nesmith, the wife of a West Palm Beach realtor, set up a Royal Air Force reception bureau in her home. She had a book filled with names and addresses of the many local families willing and happy to take us into their homes for a weekend or more and even for extended sick leave times when those needs arose. We were treated like kings, and I guess there were a number of boy-girl romances, although I don't know whether there were any RAF war brides. There would never be enough to offset the great number of American GI war brides who left England for the States after the war.

well-stocked with beer and assorted goodies. And then back to reality, back to Riddle Field, where I, shot down because of illness, was assigned to Course 26. I had returned to another place where we were thoroughly spoiled, compared with RAF standards. We were waited on in the mess hall and served real civilized food, including eggs ("How do you want 'em?"), bacon, and all kinds of fruit and dessert. Our billets at Riddle Field were very comfortable, what with the swimming pool and other unexpected delights. We even had a soda fountain on the base, tended by a lovely soft-spoken girl of no more than fifteen. She had been born in Clewiston and had never seen the Atlantic Ocean, only sixty miles away. We, who went to the beach almost every weekend, found it hard to believe.

But paradise had its downside. The Florida sun, glorious though it was, could be unforgiving for those who blithely went around as though they were at Brighton Beach in the south of England. There was a word for Clewiston, and that word was *hot*. Far from the Atlantic to the east and the Gulf to the west, this town was a stranger to those winds that gave relief to coastal towns. Oh, yes, it was hot, like 96° F, and humidity to the ceiling. The cockroaches were big enough to bully cats, and the Link trainer was a sauna from hell.*

It was a relief to escape that throbbing heat as we climbed 5,000 feet up in a PT-17 (Stearman) or AT-6 (Harvard). And of course we had a wonderful time. Our instructors, all civilian Americans, were easygoing, informal, and pleasant—direct opposites of the frustrated RAF types we had at elementary flying training school in Britain. But there was always a throat-clutching fear of being CT'd [*forced to cease training*]. Always with us was the worry about some failure in the flight or ground school phase of our training. The training was rigorous and demanding of our thought and time. We were warned to "keep our noses clean" off base and most of us did just that, pursuing our RAF pilot wings with youthful enthusiasm and an almost religious fervor. Those wings, symbols of accomplishments, were always in our dreams.

We didn't always dream—we mixed in a lot of fun during training. We flew low among the palm trees and we soared and danced among

*The Link trainer is a ground training device used in instrument-flight training. It is similar to, but more advanced than, the amusement park "driver-training" rides that simulate automobile driving.

the light cumulus clouds that ringed Palm Beach County. Buzzing Florida fields "on the deck," as American fliers called it, one of my friends killed a steer and that incident sobered us up for a little while. How we envied each senior Course as they stepped forward to receive their coveted wings. We lesser mortals marched in drill to the strains of "Always," played by the marching band. We couldn't wait for our turn at those Wings Parades of every six weeks or so. But a Wings Parade was not in the offing for me. I had been training for almost four years and it was not easy to see those magic wings slip through your grasp simply because the war was over. It was only later we realized how lucky we had been to be "the last of the many."

That grand old Florida hospitality came back a few years ago when the Embry-Riddle organization in Daytona Beach celebrated its Diamond Jubilee (1926–1986) and the "old boys" were invited to join in the festivities. The 5BFTS mustered a group to travel from Great Britain, and I drove down from Toronto with my wife. We were entertained, wined, and dined by Embry-Riddle and met our old mentor, John Paul Riddle.

I have no combat experiences to talk about. That sort of thing is the way the cards fall in life. But flying has been good to me, thanks to my days at Embry-Riddle in the 1940s.

L. J. (Jeff) Taylor

Sheffield, England

Jeff Taylor was in Course 17 at 5BFTS, Clewiston's Riddle Field. He was in a mixed class of ninety British Royal Air Force cadets and twenty U.S. Army Air Corps cadets. After the war, Taylor married a former WREN (Women's Royal Naval Service) and worked in his family's business, an electrical wholesale supply firm.

I joined the RAF as an observer [navigator] in 1941. In those days, the observer wore a half-brevet [wing] with the motif O [Observer]. Whoever wore that insignia had to bear the nickname Flying Ersehole, not something one would accept eagerly. Later they brought in an N for navigator and E for engineer to replace the dreaded O. At the end of my navigation training, I was given the option of mustering to pilot and, in mid-1943, I was posted to Florida's Riddle Field in Clewiston. We sailed solo across the North Atlantic on the *Louis Pasteur,* a former French liner that later was torpedoed and sent to the bottom by a U-boat. After arriving in Halifax, Canada, we were put on a train, went down the east coast of the United States, and stopped off for brief visits in Boston and New York. One of my early memories of this journey was leaving the train somewhere in Florida* for breakfast in a local hotel. The good ladies of the town welcomed us all in true American fashion with great generosity and affection, as was the case all the time I was in the United States. During breakfast, a pianist played some melodies, and suddenly she broke into a tune very familiar to us cadets. There was a clattering of cutlery and dishes as we all jumped to attention. I don't know who was most embarrassed, we or the pianist, when it was explained to us that the melody was a popular tune in the United States. The pianist at Hotel Sebring was playing "My Country 'Tis of Thee"—the same tune as "God Save the King," the British national anthem.

Arriving at Clewiston, we were kitted out into U.S. khaki drill uniforms, but retained our blue forage caps with the white air cadet flash. My first impression of the field was the sumptuous amount of food and the freedom after four years of war-torn Britain—its blackouts, air raids, and strict rationing. Just seeing a banana was an unbelievable novelty!

I found the primary training segment very interesting and I thoroughly enjoyed the flying. The Stearman biplane trainer was a delight to fly, but the ground instruction was long and tedious. My primary instructor was a former barnstormer pilot and was completely incredible. One day as we were flying, he said, "Hang in there and I'll show you what we used to do all over the country." He did all sorts of aerobatics that made my eyes pop in amazement and envy. He put that PT-17 through motions I never thought possible. Finally, he settled into snake-high low flying that he called "grass clipping." When we landed and got out of the trainer, the instructor slapped me on the back. "Son," he said, "the flora and fauna in Florida are not very high, and let me tell you, buddy, we were below 'em." Since the war, I have seen this fine aircraft, the Stearman, used for aerobatic displays. On one occasion, a Stearman flew by with a young woman balancing on the top wing.

*Probably Sebring, Florida—a standard train stop for each Royal Air Force class traveling from Canada to Clewiston.

When we moved to advanced from primary training, I had difficulty, but fortunately Mr. C. W. Barclay, my instructor, had the patience of Job and saw something in me that caused him to persevere until I got it right. I was eternally grateful to him for the rest of my career in the RAF because of his ability to teach. Thanks to Mr. Barclay, I qualified for my wings with distinction and a rating of excellent, which stood me in good stead in later years. One of my favorite memories takes me back to when we were scheduled for night flying and took breakfast in the PX. On this particular morning, Mr. Barclay ordered a large stack of flapjacks with a fried egg sandwiched in between each layer. Then he reached for the syrup, poured it over the flapjacks, mashed them all up, and sat back to enjoy what he had created. It was a moment of true astonishment for me. Mr. Barclay was amazing not only in the cockpit but even at the breakfast table.

We were confined to camp most of the time, but this posed no great hardship because of the facilities on base, including the swimming pool, gym, tennis courts, and cinema. On one long weekend leave, three of us decided to hitchhike to Palm Beach. The driver of a Buick 8 pulled up and offered us a lift. He turned out to be a farmer, and with the generosity typical of the Americans I met, he took us to his home and introduced us to his two beautiful daughters. Unfortunately, I was the odd man out, but the two said not to worry for they would fix me up with a friend. The father, Mr. Wells, took us all to a club for dinner and drinks, and there to meet me was the girl arranged to be my date. We all got along famously and had a great time. My new girlfriend invited me to Sunday lunch on my next leave. I duly arrived at the house and received a royal welcome from her family. We sat down to lunch, a lovely sight—southern fried chicken and round my plate a selection of small dishes containing various vegetables. Not being sure what to do, I waited to see what the family would do. A long pause followed, interrupted only by a few nervous twitches and muffled low coughs. Finally, they urged me to go ahead and eat up. It turned out that they had been waiting to see me eat in the English fashion, squashing the peas on my fork, as was our custom at home. We all saw the funny side of it, for I was the first English serviceman they had entertained.

Some days later, one of our trio fell ill and was hospitalized. Because he lost so much training time during his recovery, he was transferred to the next primary course and was unable to get leave. Feeling sympathy for him, I took it on myself to take over his girlfriend, the gorgeous Gloria Wells. Her father and I became very good friends. Once he drove me around his farm, in and around Clewiston and Belle Glade. His land stretched so far that it seemed as large as the whole of Yorkshire. It was nearing Christmas, and Mr. Wells invited us to his Palm Beach house for the festivities. At that time of year, the American cadets wore their brown winter uniforms. We knew that our RAF blue uniforms were very popular with the girls, so we decided to wear them to Palm Beach. What a fiasco! When we showed up in full RAF blue serge uniforms, the temperature must have been almost 90 degrees. So much for our vanity! We had to hightail it back to camp (this was how we now spoke in our acquired movie-cowboy style), and get into our khakis.

January 14, 1944, we were deeply saddened to learn of the deaths of two classmates, John Parry and Tony Oakley. They were flying a night navigational flight under adverse weather conditions. Both cadets were killed when their plane crashed near Fort Myers. They had both been my pals, and they are buried in Arcadia along with other British airmen who were killed in flying accidents during the war years.

On April 15, 1944, we saw the big day finally arrive. I was one of seventy-seven RAF cadets to be presented pilot wings and to partici-pate in the ceremonial parade and flyby. What a great day it was for me! Thanks to Mr. Barclay, I had the honor of piloting my plane as my girlfriend Gloria looked up from below. The Wells family all came to the ceremony, and typical of Mr. Wells, he provided a case of champagne for the festivities. Then came my final few days of leave, with tearful farewells made by all. Within a few days after graduation, we boarded a troop train headed for the long journey back to Halifax, Nova Scotia, and the United Kingdom. Imagine leaving Florida with temperatures in the high 80s and finding your-self four days later in subfreezing weather at Moncton, New Bruns-wick. I had never been so cold in my entire life.

We embarked for England on the *RMS Andes,* sailing along on the rough North Atlantic. At one point in the journey, she rolled 30 degrees each direction, making most of the boys terribly seasick. After a nine-day crossing, we arrived in Liverpool, were sent home on leave, and reported to the Harrogate aircrew reception center for our assignments. We were all split up. Some went to multiengine aircraft, some posted as instructors, and some—like me—went to fighter planes. A month later I was back on a ship, bound for the Middle East. Some weeks later we docked at Alexandria, Egypt. Here I did my operational training on Hurricanes and, subsequently, Spitfires. Then back to another ship, bound for Italy, having been posted to 111 Squadron at Rimini on the Adriatic Sea. I went on my first actual combat operation. Our duties involved low-level strafing and later dive-bombing with a 500-pound bomb slung under the fuselage. This is highly hazardous work and we were an easy target in the dive. Sure enough, the Germans got me with their deadly 88 mm flak, and I had to ditch in the River Po delta. I was pulled out of the water by two Salvation Army girls who had witnessed the action. I suffered some minor injuries—a few bits of shrapnel in the leg. A small piece worked its way out of my foot four years later.

I was soon back on flying duties until the end of hostilities in March, 1945. Our Wing (324) was transferred to Austria on occupation duties. This was where all the trouble happened—sending back the White Russians and right wing partisans to Tito. The Eighth Army was the fall guy for this operation. From Austria we went to Germany, and in autumn, 1946, I left there for England. I was demobbed [discharged] at Warrington as a Flight Lieutenant, joined 111 Squadron as a Sergeant Pilot, and left as Squadron Adjutant. Looking back now on my wartime experiences, it seems unbelievable that it all happened fifty years ago. I will never forget the kindness and generosity of all the American people I met. They were simply wonderful.

Hugh Tudor, D.F.C., A.F.C.

Limpsfield, England

Hugh Tudor was in Course 5, formed December 4, 1941, three days before Pearl Harbor, and graduated on June 17, 1942, with forty-four of fifty cadets winning their wings.

We were so young at the time of our training in Florida. I had just had my nineteenth birthday when we arrived at Clewiston at the beginning of 1941. On weekends, having completed our parade and room inspections, we were allowed out Saturday until Sunday evening. On our first weekend, about six of us (all former Manchester

University students) went to Fort Myers. The rest of the Course went off to Palm Beach, where they were royally entertained by some very generous American families. My recollection of that first weekend was that Fort Myers was a lovely town, and we thoroughly enjoyed our twenty-four hours. All in all, those Clewiston days were happy and carefree, with good companionship and marvelous weather in which to learn to fly the excellent Stearman PT-17, the BT-13, and the AT-6A. Then came a dramatic surprise. We were hitchhiking back to Clewiston (three of us in a truck) when we heard the Japanese had attacked Pearl Harbor.

My good friend Desmond Brown and I hitchhiked to Miami, where we were entertained by and became friends with a very charming couple. They always gave us a great welcome. The husband, about thirty, was killed on Easter Sunday morning, 1942, when his light aircraft collided with another plane over the heart of Miami. Desmond and I were in church with his wife when we were called out to be given this terrible news. Because of this awful tragedy, Desmond and I did not go into Miami for a few weeks, until the widow invited us to visit her. I think it provided some relief from the deep loneliness that she was experiencing.

To sum up those happy days at Clewiston, I can but comment on the tremendous kindness and hospitality that we received from everybody. In particular, we were extremely fortunate to have marvelous instructors. They were civilian pilots with very diverse experiences in flying. Some had been flying the mail, some had been barnstorming from a young age, and some had been flying school instructors for many years.

Returning to the United Kingdom, I participated in numerous operational sorties over France, Holland, and Germany. I completed forty-three operational missions,* operating at day and by night in a photoreconnaissance role using photoflashes at low altitude. On the

---

*For those operations Hugh Tudor was awarded the Distinguished Flying Cross. The citation reads in part: "This officer has participated in numerous operational sorties, the majority of which have been flown at night over France, Holland, and Germany. By day he has attacked strongly defended targets at Hamburg, Bremen, and in the Ruhr Valley. He has achieved outstanding results. Throughout his tour he has displayed high efficiency as a deputy flight commander and sets an inspiring example to all."

completion of this tour, I was promoted to Squadron Leader. I was twenty-two.

In 1949 I took part in an expedition to the Antarctic. It involved the Royal Air Force supplying two light aircraft to the first post-war Antarctic expedition, the 1949–52 Norwegian-British-Swedish Antarctic Expedition. The RAF flight comprised a squadron leader as commanding office, myself as deputy and pilot, and three noncommissioned officers of various ground trades. We left the port of London on a Norwegian sealing vessel (*Norse*) that weighed only 600 tons. Our aircraft were in crates on this cramped little ship. Someone had overlooked the fact that, to operate in the ice, she had no keel, so for twenty-eight days we "rolled" our way down to Cape Town. Then we went through the "roaring forties" and "furious fifties" [southern latitudes] and after what seemed like years arrived at the pack ice that surrounds Antarctica.

Our task was to fly about to give the ship's captain information about the ice conditions and then to find a landing point on the Great Barrier on the edge of the ice shelf. The expedition was successfully landed and after setting the various scientists ashore and seeing the establishment of the base camp, the little *Norse* set course for Cape Town and civilization, then on back home.

Hugh Tudor's Air Force Cross was awarded in 1955 in conjunction with his commanding a night-fighter squadron of Meteors and for commanding a research and development squadron during 1951–53. The Associated Press wire story of Easter Sunday, April 5, 1942, carried the story of Tudor's friend who had been killed in a two-aircraft collision over Miami. The pilot was William J. Britton, a flying instructor for Embry-Riddle Aeronautical Institute. He was survived by his widow and young son, David. Britton was the son of the former world's welterweight champion (1921–1922), Jack Britton, and brother of Bobby Britton, a well-known fighter.

Murray Cash, M.D.

Flight Surgeon, Riddle Field

The medical facility at Clewiston included ten beds with examining room, two ambulances, and medical and surgical supplies. The staff consisted of two flight surgeons and six enlisted men, rendering medical care to military (Royal Air Force and U.S. Army Air Corps) personnel. The unit conducted flying status examinations, and offered emergency medical care for Riddle Field employees and civil service employees. Because of a severe shortage of physicians in the area, the medical unit at Riddle Field provided routine medical care, in addition to emergency care, to the people of Clewiston and the surrounding area. Captain Murray Cash, M.D., was flight surgeon at Riddle Field from 1943 to 1944. In 1967 he looked back on his experiences.

My memories of Clewiston and 5BFTS are very fond ones. In establishing a rapport with the British, I had an advantage over other Americans because I had been a British subject. I was born in Toronto, Canada, and later became an American citizen. As a matter of fact, it was because of my background that I was assigned to Riddle Field.

I was medical officer from 1943 until late 1944 and I met many fine students, both British and American, and many outstanding officers and instructors. George Gibson (RAF F/Lt.) was a British officer who had trouble with fungus infections all during his stay. It got so bad that I wrote a letter to the RAF delegation, telling them I was worried about his future health. It was fairly obvious that these infections came from the swimming pool, which we tried to keep free of fungi, but in spite of our efforts the infections persisted. The British cadets were the hardest hit by the fungi. It seems that the Americans had developed some degree of immunity, giving them more protection as far as severity was concerned. But minor or major, the fungi infections responded very slowly to the drugs that we had available to us at that time.

A curious thing I noticed during my stay at Clewiston was that our venereal disease rate was very low compared to other flying training schools. Another thing I remember vividly is the appearance of the new British cadets when they first arrived. They were so pale, compared to fellow cadets who had been in Florida for some time. I would caution each new group not to try to get that "Florida sunshine" look too soon, certainly not the first week they were there. But in spite of all my warnings, I knew in advance that I could always count on treating more than a few British cadets who were in misery from trying to tan the speedy way.

The crashes, of course, were a great source of sorrow to me. I gave what treatment I could and then had the more seriously injured cadets flown to the Air Corps hospital at Fort Myers for further treatment. I don't want to remember—but I can't forget—the fatal crashes when we would walk out into the swamp areas filled with the tall saw grass, dreading what we would find. They were so young, so very young, and seemingly unafraid. A tragic waste.

We had comparatively few cases of psychological problems among the cadets—unusual in a way because they were not accustomed to being far away from home for such a long time. According to RAF custom, these few psychological problems were usually taken care of by the commanding officer. When I first came to 5BFTS I had some heated arguments with some of the civilian instructors, especially the old-timers, the weathered "fly-by-the-seat-of-my-pants" pilots. I insisted that they not fly above 10,000 feet without the use of oxygen. I remember distinctly going up with Harry Lehman, and another day with Johnny Cockrill, to prove my point, and prove it I did.

I've been back to Clewiston twice, the last time in about 1959. It's a nice, quiet sleepy town. The Clewiston Inn still stands, but it is much quieter than when all the officers and many of the civilian instructors lived there. I had dinner with Sister Downs, manager of the Inn during the war years. I'm sure that if a group of us got together over a case of scotch for a couple of days, many more memories would come forward. I'm all for it.

John Broome

Uxbridge, England

John Broome was in Course 18 at Riddle Field with both British and U.S. cadets. Following the war's end he went to work in the family's hardware store. In 1958 John started his own business, operating it until retirement in 1986.

At the outset of the war I lived with my parents and sister above my father's hardware store in a busy town fifteen miles east of London. We suffered badly from enemy air raids. They were partially responsible for the death of my sister. My mother died soon after I joined the Royal Air Force. Food was severely rationed and everything else was either scarce or unobtainable. The total blackout at night was rigorously imposed. Even the glow of a cigarette was enough to cause a "giving aid to the enemy" warning. Coming from that bleak, cheerless environment in 1943 to New York's blazing lights, well-stocked stores, and food in complete abundance was an almost unbelievable study in contrasts. The impressive skyscrapers were walls of lights from top to bottom. The sight of armed police officers was nothing less than shocking because our policemen (regrettably) do not carry guns. Anything one wanted to eat was available. The beer was quite different and it was really cold! We had to keep in mind that the traffic flowed on the "wrong" side of the street or become maimed pedestrians.

We were served white bread instead of our standard wartime loaf of brown wholemeal bread without much flavor. Coffee, not tea, was the norm. Sugar and butter were freely available on our mess tables. We were astonished to see bananas, oranges, and grapefruit—something that children under four in England had never seen. The greatest thing that deeply affected us all was the overwhelming hospitality of the American and Canadian people. They were completely generous and warmhearted.

We soon learned that train rides took days rather than hours. The train journey from Canada to Clewiston was all it took to convince us that we were in a large country. On our train journey to Clewiston we had a pleasant surprise at Lake Wales, Florida. Instead of our customary breakfast on the train, we were greeted by a fleet of cars at the railway station. All fifty of us left the train, piled into the various cars, and were taken to homes of the local townsfolk for a sumptuous breakfast. Hospitality on this scale was strange indeed to us, but what a warm feeling it generated! We got back to our cars and found in each compartment a sack of oranges—something that we had not been able to enjoy for years. But now we made up for it by eating our fill, and all the time feeling guilty that our loved ones at home were not sharing our good fortune. As our train left the Lake Wales station we waved and thanked our wonderful hosts, but not nearly enough, I'm sure. They would never know how much that glorious morning meant to a group of young strangers who were uncertain as to what future mornings would bring.

When we arrived in Clewiston we learned that we would be training with some U.S. cadets, and when we met them there were no problems from the start. We integrated immediately with friendship and goodwill. Oh, yes, we were a bit envious of their higher pay scale and we ribbed them about being filthy rich, but always in good humor and in the best of spirits. All of us, Britons and Americans, found that we had much in common from those years of 1943–44. My particular American cadet friend was Ralph Black, who lived in New York State. We lost touch with each other after graduation, but recently made contact and now exchange letters. There were so many times in the United States that I was really taken aback in delight over the kind and open attitude of the civilian population to the strangers in their midst. Helpful and friendly were not mere words with those people in Clewiston, Sarasota, Palm Beach, and other places. Two things really surprised me. No one chewed gum all the time or spoke with that pronounced Texas drawl so familiar to all film-going Britons. The Hollywood movies we saw all the time did not prepare us for America. When we first came face to face with racial segregation in action, it was a little unsettling. The separate toilet facilities and drinking fountains and the separate seating on public transportation in the southern states seemed unreal.

Hard, intense, and concentrated was the only way to describe our training. We were stretched to the limit, both mentally and physically, and it was all a necessary part of the program. We knew why we were there and knew what we had to do. And the instructors and their superiors clearly understood just how much they had to push us to get the job done.

Oh, the feeling of freedom when we were finally given a weekend pass. A marvelous escape! Palm Beach, with its lovely homes and well-ordered streets, was in sharp contrast with England's bomb-broken homes and bomb-cratered streets where we used to play. All those pictures were in my mind as I walked the avenues of Palm Beach, knowing I would soon be back in the country I would always love, a place ravaged by war through no fault of its own. Here we were, Charlie Holliday and I, two young men (our parents would call us boys) gazing wide-eyed at the beautiful large houses and immaculate gardens when a car pulled up and a charming lady invited us to call on her at an address just around the corner. She and her husband became "Mom and Pop" Feek to Charlie and me all during our stay at Riddle Field. They made us most welcome, opening their home on Chilean Avenue to us. They also gave us much encouragement in our studies to become aviators and, above all, provided close bonds of friendship when we were without our families. Upon graduation, "Mom" decided to prepare for us a traditional English meal of roast beef and Yorkshire pudding. But how to cook it? None of us knew the slightest about such things and a search for a recipe proved fruitless. "The British embassy in Washington will have the recipe for Yorkshire pudding," the ever-resourceful Mr. Feek said. "That's part of their job." After several telephone calls, the embassy staff, from the ambassador on down, put their collective talents to work and came up with a detailed recipe that was a smashing success. Mr. Feek had won the day.

After Wings Parade, I returned to the United Kingdom by way of Canada and did a short spell of flying instruction and then was assigned to the Glider Pilot Regiment flying troop-carrying gliders preparing for the Rhine crossing. I had a short period flying Halifax bombers as the war was drawing to a close and my thoughts turned to the future. I was demobbed in 1946 and married in 1947.

I did keep in touch with my American "Mom and Pop." In the 1950s they came to Europe for a holiday and were able to meet my family, and I was able to introduce them to my wife and baby son. We were not the only cadets taken under their generous, kindly wings. Cadets from earlier Courses than ours, and I am sure later Courses, were on the receiving end of their unstinting help and encouragement. I only hope that I fully expressed my appreciation and gratitude to "Mom and Pop" Feek, who had a daughter, Jean, and a son who was a U.S. Army Air Corps pilot.

John E. Lodge, D.F.C.

Crawthorne, England

John Lodge was in Course 17 at Clewiston. Formed October 4, 1943, with ninety Royal Air Force cadets and twenty U.S. Army Air Corps cadets, the class graduated April 15, 1944. Presented pilot wings were seventy-seven RAF cadets and nineteen U.S. Army cadets. At the outbreak of war Lodge was a member of the London Fire Brigade and classified as "reserved occupation" status. Under normal circumstances he could not be accepted by the armed forces. There was, however, one way to shake off the nonmilitary label: by being accepted for flying duties. Lodge had endured the blitzes long enough, fighting fires night after night. He wanted revenge against the German bombers, so he joined the RAF.

Prime Minister Winston Churchill ordered in 1940 the creation of a British Airborne Army that would consist of paratroops and glider pilots. The glider pilot units were an elite corps, all drawn from Army units, and required high standards. Having distinguished itself in several important engagements, including landings in Sicily and Normandy on D day, the Glider Pilot Regiment took part in a three-pronged attack on targets in the Netherlands. Two landings achieved their objectives, but the third, at Arnhem, failed despite valiant fighting of the troops. Many glider pilots were killed, wounded, or captured.* As a result of the Arnhem landing and the losses, there were not enough glider pilots for a full-scale operation. The Royal Air Force, on the other hand, was faced with a surplus of pilots for two reasons: (1) The German Luftwaffe, with a loss of air power by 1944, was not inflicting the damage on allied aircraft and aircrews that it had in earlier days; and (2) the various British pilot-training programs (or schemes) were turning out more pilots than actually needed. As a matter of fact, several British training fields were either closing down or consolidating. It only followed that someone in a decision-making position would act to shift the surplus RAF pilots to the glider pilot effort. The logic was clear: it would be quicker to impart rudimentary military "soldier" skills to the pilots than to embark on the long process of selecting and training soldiers to fly gliders. I was one of those "redundant" RAF pilots selected for the glider program. We carried with us a certain casual, lighthearted attitude about dress and general conduct when not flying, and this didn't draw wild applause from our new associates. In a word, they were distressed. However, we were all business when engaged in operations. The glider people knew this, but could barely tolerate the fact that we didn't carry that same image on the ground. As a result they gave us a hard time, trying to force us to conform to the proper glider pilot by-the-book model. That wasn't all bad. The discipline was probably good for us because we were to go into battle soon.

"Operation Varsity" was the name for the Rhine River crossing on

*Former Flight Lieutenant Lodge refers to the 17 September 1944 operation cheerfully code-named Market-Garden. Two American and one British airborne division parachuted into Holland to capture bridges across waterways. The almost 10,000 paratroops of the British 1st Airborne who dropped near Arnhem met fierce opposition from German troops. By the end of September, the division had lost more than two-thirds of its men. The heavy casualties resulted in great part from the inability of British tanks to cut through and join the British 1st Airborne.

24 March 1945 and came earlier than expected.* As a flight lieutenant, I was Flight Commander of "A" squadron, and an RAF flight
sergeant was my copilot. We approached for a landing with the
Horsa glider and were heavily damaged by ground fire. One soldier
was killed. An Army brigadier made the citation that led to my
award of the D.F.C. He stated that my handling of the situation,
coupled with subsequent actions on the ground, were basis for the
citation. I was pleased that it was a soldier who made the recommendation after all the hassle we had received when we joined the
Glider Regiment.

Looking back on Riddle Field, I have many happy memories, such
as the time a friend and I spent Christmas of 1943, in the home of two
elderly women school teachers. "Boys," one said, "we have saved our
gasoline ration cards to give you a tour around Miami." We drove to
so many places—an Indian village, a spot for alligator wrestling, and
then Hialeah Park. Such wonderful people!

The friendships between the American and British cadets were
wholly satisfactory and much hinged on competition in sporting
events. We learned to throw the American football, but failed completely to convert the Americans to the mysteries of cricket. There
was the occasional quip reflecting the respective merits of our Air
Forces. I recall one unkind couplet. Intentionally derisive, it went:

> *We are flying Flying Fortresses at forty-thousand feet*
> *But we only have a teeny-weenie bomb.*†

*Lodge and his fellow glider pilots and airborne troops were crowned with success
on Operation Varsity, with 832 of the 880 gliders and aircraft dispatched reaching
their designated areas. The U.S. IX Troop Carrier Command dispatched 2,046 aircraft
and gliders with equal results and few losses. The load carried comprised 14,365
troops, 109 tons of ammunition and explosives, 645 vehicles, 113 artillery weapons,
and other equipment and supplies. An hour after the drops and landings had been
completed, 237 bombers of the U.S. Eighth Air Force dropped 598 tons of additional
supplies to the airborne troops. The lower Rhine was bridged at last.

†The U.S. Army Air Corps B-17, the Flying Fortress, actually flew at about 25,000
feet on typical missions over Germany and carried a bomb load of 4,000 to 5,000
pounds—certainly not an impressive payload compared to Britain's Lancaster,
which was capable of carrying a bomb load of some 12,000 pounds well beyond the
Ruhr. It could carry a bomb load of 8,000 pounds to Berlin or any other distant
target in Germany. The Lancaster later proved itself capable of carrying the enormous 22,000-pound "Grand Slam" bomb when it became available, a feat that no
other aircraft in the world could even come close to duplicating.

Oh, yes, again looking back at Clewiston; on the evening before the Wings presentation, Course 17 was given a splendid party in Sugarland Hall. When I received my wings the next day, I had no idea I would become a decorated glider pilot with a D.F.C.

Robert Richardson

Glasgow, Scotland

Bob Richardson was in Course 3, known as the "University Flight" because all fifty cadets came from the University Air Squadrons of Oxford, Cambridge, and Edinburgh. They came to Clewiston on 7 October 1941 and started primary training. One student, Roger Crosskey, was killed on 20 January 1942 in a night-flying crash in an AT-6A Harvard. On 12 March 1942 there were forty-six Course 3 graduates awarded pilot wings. All of the graduates were commissioned as officers in the Royal Air Force. It was the only Course in 5BFTS history to have all graduates commissioned upon Wings ceremony.

When I joined 18 Squadron in December, 1942, we were based at Canrobert, a small Arab village about sixty miles east of Constantine, Algeria. We were one of four squadrons of No. 142 Wing RAF. The other squadrons were 13, 114, and 614. Our "airfield" was just a nondescript stretch of dirt—a French emergency landing strip—with a garage-size hangar. Canrobert was headquarters for the region's French administration. That was its only modest claim to fame. The big excitement among the natives came when they welcomed the periodic arrival of a French government twin-engined Caudron monoplane bringing (presumably) despatches for the administrator.

Adjoining the administration building was a gloomy-looking, fairly large prison building. Early each morning a chain gang of prisoners would be herded out by rough guards into a rickety bus where they would be driven up the mountainside to the north of the village. There was only one way to tell the prisoners from the guards: the guards carried rifles. The prisoners would work until nightfall, breaking large rocks into smaller ones, and then slowly stumble down to the bus to be transported back to what one could only hope would be a better place—or at least a place where they could welcome the night and blot out memories of the day. The prison also housed headquarters of the most frightening thugs in

North Africa—the Arab Police, with whom (quite surprisingly) we were on a warm, friendly basis. Their naturally hostile attitude may well have softened because we were much better armed than they were.

Before we arrived as replacements, the squadron had suffered an almost complete wipe-out after mounting a daylight raid on Bizerte airfield at the very northern tip of Tunisia. It was a daring, perhaps foolhardy raid designed to hamper activities of the Luftwaffe's fighter aircraft. "You will have cloud cover," the weather forecasters told the squadron at briefing. Wrong! They were supposed to have fighter escort for their obsolete Mark V Bristol Blenheims. The fighters never showed up. It was not a good day. Of the eleven aircraft that took off for the operation, not a single one got back. They did bomb and strafe the airfield, with what seemed fair results. The problem was that the German ME-109s and FW-190s chased the Blenheims and shot all of them down. Wing Commander Hughie Malcolm led the ill-fated attack and was awarded the Victoria Cross, the highest British military decoration. This award is given only for deeds of exceptional valor. It was a posthumous award for Wing Commander Malcolm. Actually, the Mark V Bristol Blenheims were obsolete in every way and never should have been allocated to frontline units. After the disastrous daylight operation at Bizerte, the four squadrons in the Wing were switched to night intruder duties, where we claimed some success.

Our change to night operations coincided with the disturbing news on 22 February 1943 that German General Erwin Rommel's second-in-command, General Jürgen von Arnim, had broken through the Kasserine Pass with sixty tanks and was at the edge of Tebessa at the western end of Algeria near the Tunisian border. Tebessa was a main market town and the linchpin of the southern flank of the Allied line. The swift German advance through the line hurled U.S. troops back through the Kasserine Pass, placing them in danger of being outflanked and attacked from the rear. Group Captain "Laurie" Sinclair was commander of our wing. He flew with his squadrons and was highly regarded and respected for being "one of us." Sinclair called a meeting in which he calmly and candidly told us that the situation was so serious that we should "be prepared to engage the enemy to the last aircraft."

The antiaircraft guns that ringed our airfield were moved up the mountainside, barrels depressed so that they could act as field guns if the tanks broke through to our positions. Sinclair said that he was not prepared to go back to daylight attacks, for the Blenheims would again be the same old sitting ducks. To get prepared for our night bombing plan, flight commanders of the four squadrons went out at dusk to mark where the enemy tanks and fuel supplies were concentrated. All crews of the Wing were briefed on what the commanders had discovered, and we each did three sorties on three successive nights, with the last sortie each night completed just as dawn was breaking. The first wave would go in with incendiary bombs to set the area alight; the second wave followed with high explosives. We went in singly, at no specified altitude, making the "party" somewhat hazardous with all the traffic coming in on the targets at different levels. You had to identify the presence of another aircraft by keeping your eyes skinned for exhaust flash in the pitch-black night. There were some near collisions, with razor-thin misses, but we managed to stop the bastards. By the end of the second night, the Guards First Armored Division had raced down from the north and chased von Arnim back through the pass. This marked the start of getting the combined German and Italian North African force bottled up in Tunisia, for by that time General Montgomery was swinging north into southern Tunisia. The enemy was caught in what looked for all the world like a cylinder, with the piston slowly moving up to "top dead center."

One night—with Kasserine Pass problems behind us—my crew and I were doing "our own thing" over Tunisia in the moonlight, flying at about 1,000 feet, looking for anything that dared to move in the hours of darkness. Peter Obruk, a Canadian, was my navigator and bomb aimer, and Phil Davies, a Welshman, was wireless operator and air gunner. We were looking for railway trains or German transports on the roads. Our Blenheim was carrying 250-pounders plus 40-pound antipersonnel bombs, nasty little things to have aboard, for they were "live" from the word go. The only thing that kept us from a departure to the next world were the caps over their noses. The 250s were fused so that (technically, at least) they would not go off in the event of a crash. This was an assumption only and one was wise to view it with a degree of skepticism. As we were cruising around, Phil came on the intercom saying he may have seen a glint of metal under trees by the side of a road. I came on down to take a look and, sure enough, there was a line of trucks drawn off the road. I circled, selected a couple of 250s, fused them and came back in for what was a real fireworks display. Around we went for another go, with Phil raking them from the turret with his twin Brownings. Down went more 250s and the 40s, turning into a wild party with petrol tanks on the trucks exploding and ground fire arcing upward in our direction, close, but not to do damage. I must admit I nearly wrote ourselves off trying to use my wing gun on a heavy machine gun emplacement and almost smashed the roof of a house in the process. I pulled the stick back, desperately trying to avoid the house and sure death. We whizzed so close to the roof that the roof tiles were ripped loose. Peter, who had a grand but terrifying view from the Blenheim's nose, came back to the cockpit. "Don't you ever do that again," he screamed, "or I'll kick you up your arse." I could well understand his feelings. Had I been in that nose cone, I can't imagine what I would have done.

Soon after that close call we moved a few miles further east, ending up at Sok-el-Arba Valley, right on the border with Tunisia. We had been re-equipped with Douglas A-20s (Bostons) and were actually nearer to the front line than were the fighter units. The twin-engine medium bomber Bostons were built in the United States and were powered by 1,600 hp Wright Cyclone engines with a maximum speed of 315 mph, pretty fast for their day. Our job was giving close support to army troops, both British and American. We were on constant readiness, all "bombed up" and ready to go from dawn to dusk. Whenever the "brown jobs" [army troops] got stuck, they called us up. We usually bombed somewhere between 6,000 to 8,000 feet in boxes of six, two vees of three in formation. Bombs were often rodded so that they made a shallow crater, but had a ripping blast effect. Even a near miss could blow the tracks off a tank. Because we were so close to the front, many of our missions lasted only twenty minutes or so, although thirty-five minutes would be closer to the mark. We would take off, our A-20s three abreast, form up in the climb, and then complete our box of six formation. Twelve aircraft from each of the two squadrons was the usual battle order; the flight commander would toss a coin to determine which squadron would come in from the north and which from the south. It was felt (or hoped) that this tactic would confuse the German antiaircraft batteries. But this was a wishful dream, not supported by facts.

The hot spot was Medjez-el-Bab, where the flak concentration was very intense and too damned accurate for comfort at six thousand feet. When our bomb doors opened, the exploding enemy shells were so close that their acrid cordite fumes of nitroglycerine, guncotton, petroleum jelly, and acetone would be sucked up into our aircraft. That was close enough. There was one operation when our aircraft limped back with what appeared to be more holes than solid fuselage. It had been pierced, punctured, and perforated to the limit. An engineering officer said that our aircraft looked like a giant colander that could be used to wash and rinse tons of vegetables.

On Easter Sunday, 23 April 1943, at about 1300 hours, a "help" call came in and we were airborne within a few minutes. The leader of our "vic"—the three-ship grouping—had engine trouble and had to go back, leaving the other pilot and me with a "blank file," so we flew "tail-end Charlies" for the rest of the formation. In the meantime our leader had gone back to base, quickly got another aircraft, and caught up with us in fierce pursuit. He then indicated that he wanted to resume the lead, but by this time we were committed to the bombing run-in. We spread out to let him come between us so that we could formate [fly in close formation] on his wing tips after he had overtaken us. I was flying No. 3 A-20 on the starboard side and looking for our leader's return from astern when there was a tremendous grinding bang, almost like an explosion from a direct antiaircraft hit. My aircraft reared up as the huge fin and rudder of the leader's Boston appeared right alongside my cockpit. Instead of coming between us, he had come up underneath us. Half of his aircraft broke off on our nose, fluttering over the top of my canopy, and he went into a spiral dive to the left, at which time I instinctively put my aircraft in the opposite direction. I was dressed only in shirt and shorts and knew that we had been "holed" from the collision because of the force of the air on my legs. As I went into the dive, Peter Obruk, my navigator, came on the intercom shouting for the speed to be reduced, as he was in danger of being blown out of his compartment in the nose. There was not one piece of perspex [Plexiglas] left, the nose had been blown away in the impact, and the four Browning machine guns in the nose were hanging out, swinging from their mounts in the breeze.

Then bad news after bad news. I could not open the bomb doors to drop the "safe" (unfused) bombs. I instructed the crew to bail out, but they chorused to the effect that they preferred to stick with me if I thought that I could get the aircraft back to base and get the tricycle undercarriage down. That was the big worry, apart from the 2,000 pounds of bombs that we would land with. Although the left engine was vibrating violently all the way back to base, we were relieved when the nose wheel came down for a very nervous landing. We learned of the damage after landing and after the bombs were removed. The entire forward section of the pilot's cockpit was thrown out of alignment to the left. The forward guns were hanging on by ragged strips of metal. The undersurface was buckled from the now nonexistent bombing panel, aft, affecting the bomb doors. It was just short of a miracle that the nose wheel leg was able to respond to the hydraulic system. A large section from a blade in the left propeller had been carved out, resulting in the violent vibrations. Reaction to the midair collision hit us only after we had landed safely and had our feet firmly on the ground. We were all ice cool in the air (at least on the surface), although I knew that if that nose wheel did not come down the crew would have to bail out and I would have to try a belly landing. We also learned that there was no recovery of the aircraft or crew that collided with us on that sad Easter Sunday.

After the midair incident, I did several missions but found that I could no longer formate. The experience had unnerved me for close contact with another aircraft. I quickly took up an offer to transfer to a group that hunted for U-boats and escorted convoys. That wasn't all. Among other things, we gave air support during the creeping, bloody, inch-by-inch Italian campaign. From Sicily we covered the Anzio beachhead stalemate, which, like Salerno, seemed like an endless struggle for Allied ground forces.

One day when we were returning from a sortie around the Anzio beachhead, I spotted an Italian sardine fleet off Trapani on the west coast of Sicily. We had been encouraged to practice depth charge attacks on flotsam and jetsam to get our hand in for the real thing, attacks on submarines. This fleet of little boats with sails was just too good to ignore. What great practice targets! Pretending that we were swooping down on a U-boat, we picked out one luckless boat in the middle of the fleet and used it as our target. Passing over it at mast height, I gave the Blenheim full throttle. The powerful slipstream from our two engines filled the sail and the little boat capsized, throwing the full crew into the water. We saw them come up from the water with fists raised in most clear "body language" with no translation necessary. "You dirty bastards" could not have been far off the mark. At the time we laughed about what we had done to some "Eyeties," whose army had recently surrendered without much of a fight. Now in reflection on the event, I do have regrets over what we did in our youthful exuberance of victory. But, then, the sardine boat crew *did* live, wet though they were , unlike some of my comrades, who did *not*. (This is the price of war.)

I was returned to the United Kingdom as a flight commander, "term expired," and posted to Air Defense of Great Britain. I flew Hurricanes and Spitfires (clipped wing Mark Vs), taking the latter over to Belgium. On the heels of VE day I applied for a transfer to Transport Command and, after some "schooling," I received captain certification and was sent out to Australia, where we operated a military air line. Our operation included service to New Guinea, Borneo, the Philippines, and Singapore. When VJ day signaled the end of the war, I had a dramatic exit. I flew my C-47 all the way back to England—and back to civilian life as well.*

Jonathan Smalley

England and Naples, Florida

After leaving the Royal Air Force, Jonathan Smalley qualified as a charter accountant (equivalent to certified public accountant in the United States). He was partner in a London firm of accountants, then became a manager of the Lima, Peru, office of the international accounting firm of Price Water- house Peat. He later returned to London and started an accounting practice in Sussex, which he operated until 1976, when he sold the practice. Smalley is a published author of both fiction and nonfiction.

Why did I join the RAF? It just happened, I suppose. I was still in school when war broke out in 1939. That morning in September, some young friends had come to our house to play tennis, something we were always doing. We took a break to watch the television set my

*Bob Richardson had a long and distinguished career with the Bank of Scotland, twenty-one years as a senior manager and a specialist in foreign business for eleven years. Richardson is a Member of the Chartered Institute of Bankers in Scotland.

↵

father had proudly purchased a few years earlier.* I'm sure my father felt a bit deflated when he invited the neighbors over to see the set. An interesting invention, they agreed, but added that the programs were an insult to their intelligence. We were watching a Mickey Mouse cartoon when the senior announcer appeared on screen and broke the news. "Television broadcasting," he said in BBC unruffled tones, "will be suspended for the duration of the present emergency." We never saw a picture on that screen again. By the time the war ended, the set had gone rusty and had to be scrapped.

The war for us really didn't get started until 1940. That summer we would sometimes pause in mid-set on the tennis court, look up and see many German planes of various types flying in formation. We would shrug our shoulders and continue playing. One afternoon an intercepting Spitfire screamed down out of control with smoke forming a dense tail behind the wild aircraft. "Bail out! Bail out!" we shouted as the plane left our sight over trees at the top of our garden. We rushed to the highest point to see where it had crashed. But there was no sign of the plane. We walked back down the hill and back to the game.

My family lived in Caterham, which we considered to be a superior suburb on the fringe of London, seventeen miles south of Westminster Bridge. Although we had air raid alerts day and night, because German aircraft constantly were flying high overhead, we rarely heard the explosion of a bomb. We did, however, see destroyed buildings when we wandered any distance afield. Incendiary bombs landed in the woods behind our garden in the middle of the night. I was the only one with enough enthusiasm to go out and extinguish them. It wasn't easy and luckily they weren't booby-trapped with explosives. We often watched the searchlights slicing knife-like across the sky in a frenzied hunt for an enemy plane impaled against a black sky. Our ears were at the bursting point from the pounding, reverberating ack-ack from Kenley aerodrome somewhere across the other side of the valley.

At dusk one evening we were visiting friends and playing tennis when we saw a red glow light up the sky to the north. The London docks were burning. At that time I used to cycle twelve miles every

*Television broadcasting was underway in a limited manner before 1939 in Great Britain.

weekday to Whitgift School. There was a rule that called for one to dismount and take cover when an air raid siren went off. I spent many an hour sitting beneath a tree and reading a book. We always had to carry a gas mask in a cardboard box slung over our shoulders on a piece of string. Every British public school had an Officer Training Corps (OTC) and I signed up. Once each week we polished our buttons, applied a white paste to our webbing, wound on our puttees, and practiced stripping down and assembling Lewis guns, weapons that had scarcely been used since World War I. I only got as far as lance corporal and jumped at the chance to volunteer for the newly formed Air Training Corps (ATC). In on the ground floor, I quickly made sergeant.

ı I wasn't the least bit interested in flying, but became quite good at aircraft navigation and, with my sergeant stripes in the ATC, I went as a matter of course with the first batch of applicants to the Air Ministry for consideration to the University Short Course Officers Training Scheme. There were few undergraduates at the universities in 1942 except for those paid for by the RAF or Royal Engineers. Successful applicants had to be "officer material" who wanted six months' exposure to university life to give them "background." After all the fitness and intelligence tests at the Ministry, there followed an interview. I stood at attention in my school blazer in front of a board of senior RAF officers and did my best to answer their questions in a manner that might impress them. When I started to leave the room as the interview ended, an officer asked, "So you want to be a Spitfire pilot and have a go at the Hun, what?" As the idea of endangering my life had very seldom occurred to me, I was a bit taken aback. But I put on my best Churchillian expression and replied, "Indeed I do, sir, by jove!"

During my final term at school and awaiting the decision of the Air Ministry, I joined the Home Guard as a despatch rider. Driving instruction was given in an old Austin saloon [luxury sedan] by an elderly chauffeur who worked for some friends of my parents. "If the guv'no reading his *Times* in the back seat knows exactly when the car starts moving, that's bad driving," he told us. We had to teach ourselves to drive motorcycles. We used camouflaged Automobile Association patrol bikes with sidecars, which flew up and cracked you in the ear if you took a left turn too fast. Driving at night with blacked-out headlights was difficult and dangerous.

Members of the Home Guard fell into two categories: very old or very young. Our sergeant, a jobbing gardener, had been fired by my father and he enjoyed getting back at me. "Toffee-nosed young bastard," he would say, and played pranks like letting the air out of my tires. It was during my Home Guard experience that I went to see a war film at the local cinema, in which several Spitfire pilots were shot down. For the first time the idea of flying made me feel nervous.

Eventually I was accepted by the Air Ministry and a letter arrived with a railway warrant with orders to report to Magdalene College, Cambridge, by 1600 hours in the autumn of 1942. I had no idea what to expect, so I simply packed a small suitcase and went along. When I reported to the porter's lodge, I was addressed as "sir" and was taken to rooms I was to share with another lad who had been chosen for the "short course." On our sitting room table was a card from the Master, inviting us to join him for sherry at six o'clock.

I bought a secondhand gown and mortar board and a pair of corduroys and settled down to university life. Thinking that physics and higher mathematics were vaguely relevant to flying, I chose to attend those lectures, only to find that I understood very little. I had specialization in modern languages at school. We put on our cadet uniforms one day each week and trained with the Cambridge University Air Squadron. One cadet each week was picked to be in charge of the entire activity, a terrifying experience. Our favorite instructor was an ex-Spitfire fighter who was always telling us to "pull your finger out" when we weren't up to standard. "Sir," asked one naive cadet, "exactly what does that expression 'take out your finger' mean?" The hard-bitten, plainspoken Spitfire pilot told the cadet exactly what the expression meant. "But, sir," the cadet asked, "why would anyone want to have his finger up his backside in the first place?"

From time to time the cadets were assembled and addressed by visiting dignitaries. The exiled King of Rumania told us that we were the cream of British youth and that the world would be grateful to us. We hadn't done anything at that point that would justify such a rosy prediction. At the end of six months we paraded in our caps and gowns and our names were recorded as members of the university. We were to report to the induction unit of the RAF in Regents Park, London, to mix in with the rest of the current recruits. We were kitted up, called "you scruffy lot of erks" [the lowest rank in

International friendships that developed at Clewiston included the local girls, who found the RAF cadets the epitome of romantic valor. Pictured here (left to right) are Tempe Ange and John Garlick (who were later married), Lois Heflin, and Clifford Mitchell. (Photo by Charles C. Ebbets)

Many American student pilots were selected to train with the RAF cadets at U.S. bases as part of the Lend-Lease Program. Two of them are pictured here (wearing helmets with goggles) near a Harvard AT-6 trainer at Riddle Field. (Photo by Charles C. Ebbets)

Cadets cram for exams. Ground training included study in meteorology, armaments, radio communications, and navigation. (Photo by Charles C. Ebbets)

Instructors wait for the "changing of the guard" from morning to afternoon sessions during primary flight training at Riddle Field. As Reed Clary, an instructor, notes, aircrews would sometimes keep a PT-17 aircraft idling so that its propwash would help keep the ferocious mosquitoes away. (Photo by Charles C. Ebbets)

the service] and drilled relentlessly by instructors who made it plain that they thought us anything but the cream of British youth. We ate one meal in the London Zoo canteen. After Regents Park, we were posted, in my case, back to Cambridge. We were billeted at one of the lesser colleges, which had gone out of business "for the duration," and taken at intervals to a small airfield where we were taught to fly Tiger Moths, bi-planes or "string-bags" as we called them. Then (joy of joys!) after five hours' instruction I managed to solo without causing any visible damage to either myself or the plane.

After a short embarkation leave, we spent a few weeks in Heaton Park, Manchester, in private homes. We tried to keep the "white flashes" on our forage caps gleaming white, but if we put them on a bedroom window ledge to dry overnight they turned a sooty gray in the morning. [The white flashes designated the wearer as an aircrew trainee.] Rumors abounded among the mass of airmen idling time away in Heaton Park, waiting transport overseas. One day we would be going to South Africa, the next day Canada, and the next, New Zealand. None of us expected to end up in the United States. Finally, without even a hint as to our destination some of us were taken by train and boarded the *Louis Pasteur* at Liverpool. A few hours later it began to zigzag its way across the Atlantic. At least, rumor had it that we were going to cross the Atlantic.

We were each issued a hammock, the main deck having been cleared of all partitions except for the lavatories. We ate, slept, and did our ablutions all on the same vast enclosed deck. At night the deck was a mass of hammocks swinging with the motion of the ship. If you were caught short and had to crawl off to a lavatory, the chance of getting back to your own hammock was slim unless you carefully counted backsides and took bearings. During the day we wandered round and round our allotted deck space to fill in time before meals. As I was passing by the trestle table [food table] I noticed a trio of disconsolate cadets sitting there with a pack of playing cards. "I say, Smalley, would you care to make a fourth at bridge?" asked one of them, whom I recognized from my Cambridge days as the young Lord Strathcarron. I knew him by sight—I had hardly been in his set. He introduced his two companions by their Christian names, then rather furtively added, "They are both peers like me, you know. Us chaps have to stick together."

At last we disembarked at a place they told us was Moncton, New

Brunswick, Canada. I was nineteen years old and had been in the RAF for almost four months. I recall very little about Moncton. I could not have been there for much more than two or three weeks, and like many other new arrivals spent some time in sick bay as a result of spending so much time in the airmen's canteen. We were offered as many glasses of creamy milk and fruit juices as we could drink, not to mention rich food, to which we were unaccustomed in wartime Britain. Our stomachs needed time to settle down.

Then, among the lists posted on the notice board, appeared the names of u/t [under training] cadets selected for number 5BFTS, Clewiston, Florida. I hardly knew what to think when I found my name on that list. We packed our kit bags and were driven off in a truck to catch a train as instructed. I can remember changing trains only twice, first at Grand Central Station in New York, where we waited in a group for the Miami train. We were there through the evening commuter rush with a constant bustle of passengers jostling past us, most of them cheerily calling out some sort of greeting. We were quite unaccustomed to such bonhomie. Feeling that we were some sort of sideshow, we started singing together: "Roll Out the Barrel," "I Belong to Glasgow," and "Land of Our Fathers." We all enjoyed a good sing-along. We could see the floodlit dome of the Capitol in Washington, D.C. and I think we must have changed trains in Jacksonville, Florida. We continued on through Florida during the night and, to get some air, opened all the windows. As a result, we were covered with soot from the steam engine. When we stopped at Sebring, Florida, in the morning we were welcomed by a reception committee who had arranged a table laden with huge jugs of orange juice. They welcomed us, filled us with juice, and then we chugged off again, thinking what nice people Floridians must be.

I don't remember when we left the train and were driven off to Riddle Field, Clewiston. My first impression of the field was that the sun was very bright and it was very hot. We were allocated beds in the barracks and issued American lightweight uniforms and underwear. Only our blue forage caps with their white flashes identified us as British cadets. There was an ice cream bar in the airmen's recreation room that dazzled us with its variety of goodies, and there was a jukebox incessantly blaring out tunes. Best of all was the large swimming pool. It was like arriving at a camp for summer holiday, except at home we didn't know about such places.

BFTS graduates await departure from Clewiston. They faced an uncertain fate in heading off to further training and eventual battle overseas. Almost 71,000 British fliers lost their lives in combat during World War II. (Photo by Charles C. Ebbets)

Air Marshal D. S. C. Evill presents an RAF cadet with his wings. Of the 1,800 cadets who entered the six-month program, 1,400 graduated. (Photo by Charles C. Ebbets)

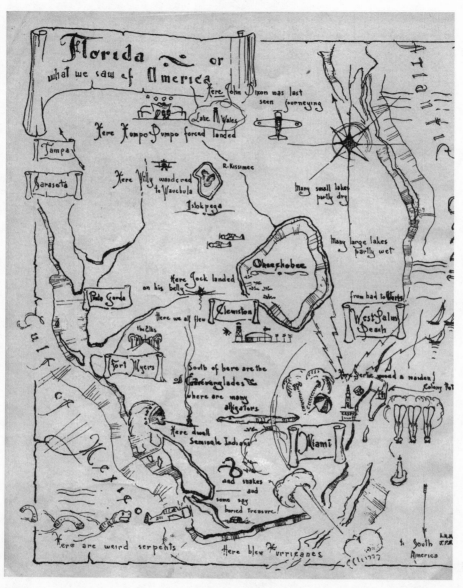

A whimsical Brit's-eye view of Florida. Reprinted from *No. 5 British Flying Training School First Anniversary: July 1941–1942.*

Right from the outset of my training in Course 16, I knew I was not a born pilot. I survived about one hundred hours of solo flying only by being scared and very careful. Flying the Stearman, I acquired reasonable competency in the primary phase of training and was passed on to Harvards. Perhaps the worst experience I suffered was landing an aircraft at night on a makeshift flare path, at the end of which stood knots of instructors and fellow pupils. They were soaked in citronella to ward off clouds of outsized mosquitoes, such as I have only seen in Florida. Sitting all alone in that Harvard and telling myself that I could land without incident between those dim, flickering flares didn't really help. The entirely unpredictable space of time that elapsed between the moment I first expected to touch down and the actual moment of impact was usually long and terrifying.

About three weeks before I was due to be given my wings, I was thrown off Course 16 and washed out of pilot training. By the time I was due to graduate, 5 February 1944, there was a surplus of RAF pilots in Europe. The huge training effort had been too successful, it seemed. Two experienced, battle-hardened fighter pilots were sent to Clewiston (and no doubt to other BFTS) to cull out the weakest performers and stop the flow of pilots. The check pilot who tested me had a hard, flinty look on his face which relaxed into a grim smirk when I made my first mistake. He suddenly killed the engine and told me to demonstrate the routine for forced landings. My descent was aimed straight into some murky swamp, not toward solid ground. Even my tormentor, direct from Europe, could spot the difference. His pleasure quickly turned to apprehension only at the very last second, when he pushed the throttle and cleared the foliage by inches. When we landed back at the field, the wheel brakes were either on or stuck and the plane almost did a ground loop. The hot, mean-spirited check pilot jumped out of the aircraft, barely concealing his mixture of relief and joy at not being down in some swampy home for snakes. Acting entirely unpleasant, he said, "I'll turn my report in to the CO," in as gruff a manner as possible. Before the day was over, I was off the Course. Suddenly I no longer "belonged." I felt quite numb and lost.

That sad last weekend in Palm Beach was one of loneliness bordering on despair that only the very young can know—the despair that comes with looking ahead and seeing nothing. When Sunday came around, the day when we would take the truck back to Clewiston, I decided to stay an extra night, and why not? I was no longer part of the Course, no longer needed and, worse, no longer wanted. No one would be calling the name of a nonperson.

One of my friends and I had gone to Palm Beach for the weekend to drown my sorrows. He felt sorry for me and when he had to get the truck back to the base suggested I take his girlfriend, Lulie, to a nightclub for dancing. Lulie was very nice to me, keeping the conversation on a single subject: What it would be like to be married to my friend and live in England. As it happened she never found out. She married someone else and went to live in Virginia.

I was posted back to Canada on a long and lonely train ride. When I changed trains at Haines City, I saw a piano in the waiting room. I put my belongings on a bench, sat down at the piano and played a few numbers—probably blues. Soon a few people gathered around and gave me encouragement to play more. A young couple asked me to play at their wedding ceremony, which was to be the following day, but I explained that I had a train to catch. The next connection was somewhere in Pennsylvania. I had several hours to wait for a connection. A fellow passenger, an American soldier whose family lived nearby, invited me to join his family for supper and I gratefully accepted. His friendly family couldn't have been nicer and, after supper, they had me lie down for a few hours on a spare bed. That would never have happened in England.

On to Canada for a new assignment and a new challenge that showed me that so-called failure could be a start for success. (I don't mean this to sound like one of those "you can do it if you try" platitudes.) I eventually came near the top of my class at air navigation training at London, Ontario, and was commissioned pilot officer navigator upon graduation in Transport Command. Most of the chaps in my course at Clewiston who got their pilot wings rarely flew again, but I continued flying.*

*Smalley's pilot training class (Course 16) had eighty-three Royal Air Force cadets in the beginning on July 23, 1943. At the completion of the Course on February 5, 1944, there were sixty-five who graduated at Wings ceremony. Smalley certainly did continue flying, even though he did not win pilot wings. He was posted to the RAF Southeast Asia Air Forces in early 1945 and saw much active service, navigating transport aircraft under difficult, and often hazardous conditions. He left the service in late 1946.

Leslie M. Knibbs

Sorrento, British Columbia, Canada

Les Knibbs was a Course 22 Royal Air Force cadet at Riddle Field. He served in the RAF until 1947 and went back to his civilian governmental position in the treasurer's department of the Slough Borough. In December, 1948 he resigned from the borough and emigrated to Canada, where he has lived ever since.

I became a member of the RAF volunteer reserve in late 1942 and was placed on deferred service until June, 1943, when I was told to report to the Air Crew Receiving Centre in St. John's Wood, London. We were kitted out with uniforms, marched here and there for inoculations, physical examinations, dental checks and the like. After two weeks and a half-dozen blisters on each foot, I was posted to an Initial Training Wing at Newquay in Cornwall—march, march, march, and then school work and more school work.

At the end of three months there were exams; those who passed (I did) were sent to a pre-elementary flying training school for flight evaluation. I soloed in five hours in a Tiger Moth, which was some sort of a record at that time, and was posted to Scarborough in Yorkshire for a post–Initial Training Wing. This was a new RAF idea—to have us brush up on our school work. Then I spent a few weeks on a Bomber Command airfield in Lincolnshire. A small flight of aircrew under training were shown all the various activities of a working airfield, including attending several funerals. We had several flights on Lancaster bombers for orientation purposes.

In April of 1944 I was posted to Moncton, New Brunswick, Canada, to wait to be sent to one of the Canadian flying schools, but later in the year I struck it lucky. Orders came through for more than a hundred of us to go to 5BFTS in Clewiston, Florida. We were thrilled. The train journey was indeed "a journey and a half"—nothing like the short distances in Britain. The United States had huge trains by our standards and such comfortable sleeping bunks. We went through Boston (most British people know something about Boston!), New York, Washington, and all down the East Coast to the Georgia swamplands, where the train was held up because the train ahead broke down. When we were going through Florida, the train stopped at a convenient spot—right alongside a tempting orange grove that had a magnet-like effect on the RAF cadets. Many hopped off the train to pick a big Florida orange right off the tree—really something to write home about!

We found that the work at Riddle Field was hard and we simply had to perform. If we didn't, there was a simple answer: we would be washed out and whisked back to Moncton for reclassification to navigator, bomb aimer, or air gunner. Our primary training segment introduced us to the famous Stearman, the PT-17, a perfect machine for learning the basics of flying—a stable aircraft indeed. The middle course brought on the Harvard, the AT-6, an aircraft that seemed enormous after the Stearman. Today it would be like going from a Toyota Corolla to a Bentley. The AT-6 brought us into the world of the variable-pitch prop, the retractable undercart [landing gear], the flaps, and a more powerful motor. We learned instrument flying and chalked up many hours on aerobatics—a lot of fun. The senior course (we wore red flashes on our caps instead of white) was spent in honing the skills we learned in the middle course. We also spent a number of hours in tight formation flying, keeping both eyes on the leader, and not looking over the countryside while in close formation. We did a fair amount of air-to-air and air-to-ground gunnery, using the photo guns, of course. Every so often we would receive a summons to report to the flight line to find out if we were progressing according to schedule. Then came the final exams and flying tests. We had a total of about 220 hours in the log book, but I had slightly more because my instructor had been promoted to flight commander and was allowed only one pupil—I was it.

Even though training was tough, hard work, we did have free time on some weekends. The Nesmiths of West Palm Beach set up an organization to allow us to be billeted by kind people in Palm Beach or West Palm Beach, and many friendships were made. I understood that Mrs. Ira Nesmith was honored by the British government for her part in the war effort, but I never found out what honor she received. I do know she was given special gasoline rationing from the U.S. government for her work with the cadets. I wrote to the Nesmiths several times after the war and always received a reply. It was not possible to take a walk in West Palm Beach because so many people would stop their cars and offer lifts. One of the large hotels in West Palm offered us rooms over the Christmas holidays with food included and a gift for each cadet on Christmas morning—all at no charge.

There were no fatal accidents during our course, though we saw the usual amount of accidents such as landing with the undercart up, ground loops, and collisions while taxiing. I had a couple of near misses. I was stooging [performing nonoperational flying] around the airfield, solo at that time, when the top cylinder pot on my PT-17 decided to detach itself from the rest of the engine. I made a rather flat descent to the grass and landed safely in spite of a couple gallons of oil all over the plane and the pilot. A second incident happened when I was about to land an AT-6 and was about twelve feet up from the runway when another AT-6 landed on top of me. The prop, about six inches from my head, had chewed halfway through the crash bar. Had anyone been in the back seat there would certainly have been a fatality. The pupil in the plane that landed on top was a very good friend and was flying dual with his instructor. He had a lot of explaining to do. The Commanding Officer was happy there were no casualties, but a bit miffed about the damage to two brand new AT-6s. After the usual Court of Enquiry, I was taken up by my instructor just for a ride. When we landed, he did a ground loop, luckily with nothing in the way. For a brief moment I thought that it might be wise to quit the flying business, but, then, we were close to Wings Day, and the thought quickly left my mind.

The long-awaited Wings ceremony was held on 31 March 1945, a little over a month before the war with Germany was to end. Course 22 had started with one hundred RAF cadets on 29 August 1944. It finished with sixty-two. Forty-one graduates became Sergeant Pilots and twenty-one were commissioned as Pilot Officers. Riding the train back to Moncton, New Brunswick, we all were eager to be posted where we could put all our training to good use. We had a day in New York between trains and did plenty of sightseeing. As we strolled in midtown Manhattan, many New Yorkers called out to us and wished us luck. When asked if we were pilots, we said yes without explaining that we had just graduated from Riddle Field.

We stayed but a short while in Canada and then traveled back to Britain. I was told I was in Transport Command and would fly DC-3s and I started a conversion course. But then the war in Europe ground to a halt, as did my future training. I was then sent to an airfield in the north of England. I carried commonwealth navigators and bomb aimers on cross-country flights to learn map reading, European style. Here again, this course was discontinued after three months and most surplus aircrews were reclassified to clerks, radio or radar technicians, or any one of a dozen trades. No longer flying, I became a base accounts clerk, an AC2, but still with aircrew rank and badges. There were three hundred flight sergeants and warrant officers at base accounts, housed in comfortable apartments. In 1946 all of them were made sergeants, but the aircrew pay and allowances were left at the level of our old rank. The commissioned aircrew ranks were frozen unless they signed on for another five years. Promotion was very slow for them—in fact it was nonexistent. The same thing applied to the noncommissioned aircrew who signed up for the long term. I left the Royal Air Force in June, 1947.

I consider myself fortunate to have been trained in Florida. The great kindness of the people and the way they accepted us was most exceptional, because we were so accustomed to the well-known British reserve. Here we were surrounded by people so open and fun-loving, even in time of war. What a change for us! I can still see in my mind's eye some of the folks that I stayed with and those who invited RAF cadets to parties.

Reputation has it that the training course at a U.S. flight training school was tougher than the course at a British or Canadian school. Maybe it's true—it's hard to know. The standards were high indeed at 5BFTS, Riddle Field, due to the caliber of the flying instructors and ground instructors. My only regret is that I went through years of training, passed scads of examinations, and went for years with low pay, but was never able to do the job I was fully trained to do. That was the only low point in my RAF career.

Harold Charles Skerman

Cheam, England

Harold Skerman was a Royal Air Force cadet in Ponca City, Oklahoma, before being posted to Riddle Field at Clewiston, Florida. When the war ended he joined the family business, which produced packaging machinery for the food and pharmaceutical industries. Skerman returned to his love of flying in 1963 and tested for his private license. At various times his company owned a Beach Musketeer, a Piper Twin Commanche, and a Beach Barron. The planes were used for business and pleasure throughout the UK,

Europe, and the Middle East. Skerman and his wife are frequent visitors to Florida.

It was my ambition to become a Spitfire pilot, but I never made it, even though I did get my wings. I joined the RAF in 1943 at age nineteen and so my active service career was very limited. I had a normal schoolboy's life, living peacefully with my family, never doing anything really exciting or traveling very far. Suddenly I was in the RAF with one burning ambition: to learn to fly and wear those pilot's wings. Today I think back on those times and realize how interesting they were.

Everything that happened day by day was a new experience for a boy who found himself involved in activities he had only dreamed of a short time earlier. The rigors of training were no great worry. All my friends and I wanted to do was learn all we could about flying and then *fly.* Our highs and lows centered around whether we had passed each examination with good marks. The continual traveling to new places in England opened my eyes to a larger world. Who could have wished for a better experience than boarding the *Queen Mary* and heading for New York? Being so far away from home and family was strange, but I had little time to think about such matters. There was so much to do, and I clearly remember how kind the Americans and Canadians were to us when we were either training or on leave.

My training in the United States carried me almost from one end of the country to the other. When we arrived in New York, we received a greeting that I will never forget as we left the ship.

◡: The Yanks ∿

They ranked high on the list—perhaps the highest—among the flying cadet cadre in the U.S. Army Air Corps when it came to background, training, and raw potential. They were the 125 American student pilots who were handpicked to train alongside Royal Air Force cadets. Starting with Course 12 (November 12, 1942) a few of these U.S. Army cadets were intermixed in each course with the British trainees.

The Americans were carefully selected on the basis of test scores and personality profiles. Virtually all had some college or university credits before entering the service. A fair percentage had received four-year degrees and even more were within a few academic hours of receiving a degree. On the average, they were somewhat older than the RAF cadets. Finally, the 125 Americans had chalked up some flying training as civilians, limited though the actual number of hours may have been.

Several theories have been advanced as to why a sprinkling of American cadets was included in a British flying training curriculum, even as a token gesture. Would their presence in the British courses help the RAF cadets fit more quickly into a new culture? Interesting, but simply not true. The RAF cadets were fitting in quite well without any assistance. Rumors over the anticipated appearance of the "Yankees" spread throughout 5BFTS as British cadets added layer after layer to each unverified report that gave expanded, and frequently mystifying reasons why the Americans were coming. A few British cadets suggested that the U.S. Army Air Corps cadets were brought in to "Americanize" the British training plan by introducing the dreaded West Point system of discipline adhered to by the U.S. Army Air Corps. (See Carlstrom Field, p. 5). All rumors were quickly squelched when RAF Wing Commander Thomas O. Prickett and Captain Thomas E. Persinger, USAAC, issued a joint statement clearing up the wild speculations concerning the arrival of the American "brown shoe" flying cadets. Prickett and Persinger succinctly pointed out that the decision to absorb a few U.S. cadets into each Course did not herald any change in the structure of the British flying training syllabus. The sole reason for the new approach contained no elements of high drama. It was done merely to balance the Lend-Lease Act account between Great Britain and the United States.

When the first seventeen Americans moved in on November 12, 1943, to take their places with eighty-three RAF cadets in Course 12 at Riddle Field there was instant acceptance. On the base were 200 RAF cadets in Courses 8 through 11 working their way through basic and advanced. Course 8 had graduated thirty-nine of its original fifty cadets on November 11, the day before the Anglo-American Course 12 cadets arrived.

November 12, 1943, fell on a Friday, a free weekend that gave rise to a new alliance and a splendid excuse for a celebration. The Yanks were first introduced to Werts, a favorite Clewiston tavern that had found its way into RAF rhyme "Whiskey and Soda at Werts" had become a favorite British byword since Riddle Field opened. As the night went on, one group decided to transport some of the joy and newfound fellowship to Charley Steele's Tavern at Punta Gorda, about eighty miles from Clewiston and a favorite spot for the British cadets. Three softhearted Werts patrons drove the cadets to the Charlotte Harbor town, despite severe gasoline rationing to discourage joyriding. (The cadets—British and American—who recently had been paid, passed around the hat to compensate their drivers quite generously for the taxi service.) Before Charley Steele closed his tavern on Friday, he arranged with local residents to house the ten or twelve cadets for the night. Steele received reports the following day from his friends who had opened their doors to the cadets. "They were a great bunch of kids," one host said, "full of harmless hell, but behaving as real gentlemen." The Punta Gorda contingent, who had left the main body of celebrants at Werts in Clewiston, hitchhiked to the Sarasota Hotel on Main Street. After taking showers and resting for a while, the cadets set out in the chilly (for Florida) 60 degree afternoon in search of a place to have lunch. "We walked around town for a bit," said RAF cadet Tom Flowers, "then we found a lunch counter spot [probably Badger's or Walgreens] and had milkshakes and sandwiches." The cadets went to the Army and Navy Club and struck up conversations with the available, willing girls, who suggested they all go to the Tropical Lounge on lower Main Street to hear pianist Carl Springer. When Springer spotted the cadets he played the "Air Force Song" and "There'll Always Be an England." "The place closed down when the pianist ran out of steam," said RAF cadet Tim Lane. "Besides that, we were all getting hoarse from singing and a couple of lads were a bit one-over-the-eight [feeling no pain]. We went back to the hotel, slept until Sunday noon, and hitchhiked back to Clewiston."

Short-lived though it was, the mixed Courses idea turned out to be an innovative change that was a "natural." The Americans and British took something from each other during the seven-Course span and learned to appreciate both their similarities and differences. The blending of cadets ended after completion of Course 18 on June 17, 1944. The U.S. Army Air Corps, facing a surplus of pilots, withdrew U.S. cadets from all BFTSs.

The seven Brit-Yank courses were broken down as follows: Courses 12, 14, 15, and 16 each had seventeen USAAF cadets and eighty-three RAF cadets. Course 13 also had seventeen USAAF students, but only seventy RAF cadets. Courses 17 and 18 each had twenty USAAF students and ninety RAF cadets. Sixteen of the 125 American cadets failed to get their wings, a 12.8 percent washout factor. Among the 582 British cadets in the six courses, there were 121 who washed out, a 20.8 percent washout factor. The low washout rate for the 125 Americans simply confirmed that they were the "best of the best" among U.S. Army Air Corps cadets. They each had a leg up because of prior flying training, a high degree of mechanical skills, and maturity. But when it came to an even playing field, without preselection of natural talents and experience, the average Army Air Corps washout rate and British BFTS washout rate were almost identical—from 20 percent to 25 percent. RAF cadets in the Arnold Plan (see Carlstrom Field, p. 5) had a washout rate of 40 percent or more, double that of cadets in the British flying training schools or U.S. Air Corps cadets.

Most of the U.S. cadets who won their wings (both American and RAF) at 5BTFS went on to multiengine training and assignment to the Air Transport Command. They frequently distinguished

themselves in all theaters of war, moving supplies and troops under hazardous conditions. In addition, they ferried aircraft of all
types, fighters to bombers, throughout the world. Many of them
became airline pilots after the war and others remained in the Army
Air Force until retirement.

Fast friendships were forged during the seven Courses, which
saw a small group of U.S. cadets and a much larger number of
British cadets sharing common experiences, intense ground training, vital hours in the air, great joys and deep sorrows. Thirteen of
the twenty-three graves at Arcadia's Oak Ridge Cemetery are those
of RAF cadets who were classmates and friends of the young American men who trained with the British during that brief sixteen-
month period.

In 1994 there were forty former U.S. cadets who trained with the
British at Riddle Field listed as members of the Britain-based 5BFTS
Association. After more than fifty years, the bonds that stretch
across the Atlantic are stronger than ever. The American members
and their Courses:

Course 12:   Doyle C. Alexander, John B. Gillette, Peter J.
             Lazarra, E. Sinks McLarty, Fred R. Renshaw,
             Robert R. Rissman, Richard P. Schmidt, Blaine H.
             Schultz, Otis O. Skubal, William A. Slade, Clifford
             L. Suhm, and Charles G. Weber

Course 13   Robert F. Agne, Alfred A. Greenberg, Col. Samuel
             R. Huston, Bromfield L. Ridley, and Richard J.
             Warner

Course 14:   John W. Cook and John D. Roy Jr.

Course 16:   Paul D. Danforth, Joseph W. Harpham, Robert G.
             Royce, Norman H. Stevens, Tom Masano, and
             Harry J. Parrish

Course 17:   Kenneth E. Mills, J. T. Moore Jr. M.D., Maj.
             Charles A. Neyhart, Ormond Roberts, and Arthur
             Rushworth

Course 18:   John C. Alberts, Paul E. Ardley, Ralph D. Black,
             Edward Donowick, Harold B. Kinison, Jim Mead,
             Douglas Moore, Charles A. Sweet, and Walter B.
             Thomas Jr.

John Clark Alberts

North Barrington, Illinois

John Alberts was in a group of U.S. Army Air Corps cadets specially selected to train with British cadets at Riddle Field. The experimental program started in Course 12, November 12, 1942, and was discontinued at completion of Course 18 on June 17, 1944, primarily because the Air Corps was approaching a surplus of pilots and the RAF wanted an increased intake of British cadets. The seven "mixed courses" were hailed as successes by both British and U.S. officials.

In recent years much attention has been paid to our World War II experiences. All the men I trained and flew with yearn to recapture at least a shred of those days; there is a clinging, a reluctance to loosen one's fingers, and watch part of us slowly drift into dusty pages of history. I'm inclined to think that it is not mere nostalgia, nor a longing for lost youth and great adventure (although I'm sure that's part of it) that compels us to hold on to the memories of those days. Can this be the answer: We are in our seventies and eighties and our families are raised and departed from us physically; we now have a yen to renew old friendships and relive some of the most important days of any generation and we have the time to indulge ourselves.

Back when the talk began that led to the training of British fliers in the United States, we were "neutral," although under the Lend-Lease Act we did supply war materiel and ply the Atlantic with ships to aid Great Britain. A schoolmate of mine from Taft School, Watertown, Connecticut, went down in Icelandic waters on one of those missions. He was Howard "Bud" Wade, one of America's first casualties in a yet undeclared war.

Our prewar participation made possible the training of British Royal Air Force aircrews (something that proved to be a good move) and someone came up with the brilliant idea of mixing British and U.S. cadets in flight training. A small percentage of the U.S. cadets would be enrolled in the Courses at the British flying training schools, promoting "hands across the sea" and all that. I probably would have been cool to that idea at one time because my Irish forebears didn't instill in me any great love for the English. But, then, as a student at Taft I had a good friend, an English exchange student named Alex Havey with whom I spent much time. He visited our home in Plainfield, Connecticut, several times over weekends. So opinions do change (as mine did) and besides, the British were opposing Hitler.

All of the preliminaries of the mixed Brit-Yank Courses found me at Maxwell Army Air Force Base, Montgomery, Alabama. We were due to ship out to various training fields when the word was passed that twenty men were needed to fill a requirement at a British training base. Although not evident at the outset, only those U.S. cadets with some college credits plus previous flying time would be considered. The profile seemed to fit me right down the line. I had two years of college when I enlisted as an aviation cadet and had completed a civilian pilot training program. My name was on the list when they called out the chosen twenty and I quickly learned that all of us had college credits and some previous flight experience. We asked ourselves why the Air Corps had insisted on the college and flight credentials. Was it done to make sure we put up a good show for the Brits or merely to make us even with them at the start? The British cadets on average beat us in the classroom ground courses, but not in actual flying.

When twenty of us arrived one late afternoon, we were quartered with the RAF cadets, had our dinner, and then went to sleep on cots in a large common room. We were dead to the world, sound asleep at 5:30 A.M. (oh, it seemed earlier) when there was a cannon-like roar that kept repeating, "Wakie! Wakie! Wakie! Show a leg! Show a leg!" Issuing the ear-shattering repetitive bellow was an immense red-faced British Leading Air Craftsman who stood, we thought, seven feet tall or more. We were quite off the mark. His official height was only six feet, four inches. And what in the hell was this "show your leg" stuff all about?* I remember the first time I heard someone use the expression "you've had it." It was English mess

---

*The term "show your leg" dates back to the glory days of the British fleet. When a man-o'-war came into port, the ordinary sailors were permitted to bring girlfriends on board and they could sleep together in hammocks.When standby watch was called out, the bosun went up and down the hammock line shouting, "Wakie, wakie, show a leg." If a slender, smooth leg came out, all was well, and the owner and friend would not be disturbed. If a hairy, brawny leg came out, he was brought out for duty. All sorts of tradition, or as the English say, "bags" of it.

sergeant lingo for, "Buddy, you're too late for chow. The mess hall's closed."

And so to Riddle Field. We spent seven months in training using PT-17 [Stearman] biplanes in primary and AT-6 [North American] low wing monoplanes called Texans by Yanks and Harvards by Brits.

After seven months there were four out of the original twenty U.S. cadets who washed out. Sixteen of us graduated and fifteen became second lieutenants in the U.S. Army Air Corps. The seventeenth American who graduated was denied a commission and made flight officer, something between enlisted and commissioned, signified by a little gold bar with a blue mark. The big guns found out that he had married a girl during the time he was a cadet. This was strictly verboten, but too much money had been invested in him to wash him out so they held back his second looey bar.

It is worth noting that after graduation only eleven RAF cadets in Course 18 were commissioned pilot officers, while the other RAF graduates who won pilot wings were made pilot sergeants. It was felt by most Americans and a goodly number of Britons in the class, rightly or wrongly, that the commissions were awarded on the basis of family position or old school ties. This was a virtual certainty for any "best" cadet in any class at Riddle. The class system, we said, "was still functioning." There was one notable exception. RAF Cadet Kenneth Rudd, chosen by instructors and officers to be "best all-around cadet" in his class, was commissioned.

Former U.S. cadet Alberts, without having actual 5BFTS records on hand and looking back fifty years, was extremely close in citing the graduation figures and number of commissions awarded for British and American cadets. He was right on the target in stating that four of the twenty U.S. cadets in Course 18 did wash out and that commissions were awarded to fifteen of the American graduates and a flying officer rating to the sixteenth graduate.

He was close also on the British graduates. There were ninety RAF cadets who started in Course 18, of whom eighty-two graduated with wings. Fifteen of the graduates were commissioned pilot-officers, with the remaining sixty-seven graduating as pilot sergeants. (See page 146 for Courses 12–18 statistics.)

Was there resentment about the fact that over three-quarters of the RAF pilots were made pilot sergeants while other graduates in the same Course received pilot-officer commissions? A qualified yes unless all the rules of human nature had been completely overhauled.

The official policy of the RAF high command regarding commissions remained: "A commission is granted in recognition of character, intelligence (as distinct from academic qualifications), and capacity to lead, command, and set a worthy example. Many aircrews, though quite capable of performing their duties adequately, have no officer qualities."

Taking exception to the British Royal Air Force position, and not unexpectedly, was the Royal Canadian Air Force, which clamored for commission status for all aircrew members. Canada spelled it out, but could not sway the Royal Air Force: "All aircrew members should be given commissions. It is not right or proper that a noncommissioned officer (NCO) should be expected to perform the same duties as a commissioned officer but without the rank that goes with those responsibilities."

In some ways the British policy on aircrew commissions made sense, although General Jimmy Doolittle was reported to have said, "it is something like playing tennis without a net."

RAF rules stated that you could be a commissioned wireless operator, flight engineer, or gunner as long as you possessed those hard-to-define "leadership qualities." You could wash out of pilot training, go on to bomb aimer or wireless operator training and come out with a commission. You might find yourself on a crew with a sergeant pilot who had won his wings at the same time you were washed out. It did happen, but it never seemed to cut into the resolve of the RAF aircrews when it came to winning the war that had to be won. Bitch all you want, but keep on doing your job.

Something else must be mentioned. Many RAF sergeant pilots were commissioned soon after going into operations against the enemy. Others, of course, were killed soon after going into operations.

The U.S. Army Air Corps had little or no trouble with a clear-cut policy on aircrew commissions. Even though at one time there were Air Corps pilots with sergeant stripes, that went out the window soon after Pearl Harbor. The Air Corps then began awarding new pilots second lieutenant rank or, for some mysterious reason, the dreaded flying officer designation. This, too, changed and soon all new pilots graduated as second lieutenants.

With sergeant pilots and flying officer ratings made a thing of the past, all pilots, bombardiers, and navigators in the Army Air Corps were commissioned—period. Radio-gunners, flight engineer-gunners, and regular gunners held enlisted rankings, usually staff or technical sergeant. There *was never a move to provide commissions for enlisted crew members because the length of training for a pilot, bombardier, or navigator was so much longer and more intense. Technical Sergeant Edwin "Red" Bain, a veteran of the daring Tokyo bombing mission, took a tolerant view of commissioned officers. Dealing cards at a poker game in a Sardinia-based Marauder bomber squadron noncom club, Bain said, "You've got to watch out for those kids," in his soft North Carolina drawl. "They need a lot of help."*

Blaine H. Schultz

South Milwaukee, Wisconsin

Blaine H. Schultz chalked up more than 1,700 hours as a pilot during his military career (1942–45), flying with the Air Transport Command. He flew everything from fighters to multiengine aircraft. Schultz returned to college after the war and received his electrical engineering degree from North Dakota State College, Fargo, in 1947. He went to work for Line Material Industries, a McGraw-Edison Company, and in 1959 became an engineering management executive, a position he held for twenty years.

I was a farm boy in North Dakota, studying electrical engineering (1941–42) at the North Dakota State College in Fargo. In my sophomore year, I took primary and secondary pilot training. Before taking the secondary course, it was required I sign an affidavit that I would join the U.S. Army Air Corps in the event of an emergency. I had no problem with that, of course, but my parents did have great reservations; finally, they reluctantly consented.

In August, 1942, on my way to enroll in the fall semester at college, I stopped to check the mailbox. There were a couple of farm supply catalogs, a letter to my father, and a governmental-looking envelope addressed to me. I opened it at the mailbox to find a terse message and a railroad ticket to Nashville, Tennessee. The message told me to leave that very night for the "Classification Center" in Nashville. My plans for college were, as they say today, put on hold. I was sent to Maxwell Field, Alabama, for preflight training. The word came down that volunteers with previous flying experience were needed to attend the British Flying Training School in Clewiston, Florida. In return for volunteering, we were told that we would be placed in the noncombat Air Transport Command. That seemed to be a good idea at the time. I was one of seventeen

U.S. cadets in Course 12, along with eighty-three British cadets assigned to 5BFTS in Clewiston, Florida.

When it comes to training, every cadet at Clewiston must remember the low flying episodes—and I do mean *low*. We would fly down the canals so low that you couldn't look over the banks. I recall a man fishing from a boat in a canal. We came down in a deep dive and saw him lie down in the bottom of the boat as we roared over. We chased the Brahman cattle all over the landscape. The British cadets in the air would play chicken with cars on the highway. One RAF flier flew so low he touched the road surface, bending the tips of his prop. To avoid being caught, he landed with his wheels up, thereby really bending (or rebending) his prop. "Oops," he reportedly said, "I forgot to lower my undercarriage." I can't vouch for that story, but it could have happened.

The British cadets came from a wide variety of backgrounds. I was told that the earlier courses had mostly upper-class cadets and that later the RAF was accepting trainees from the working classes. This may have been accurate, judging from the many speech patterns and accents. One of the Brits in Course 11 had been a pianist for a band and frequently entertained us in the canteen, where we bought Cokes, ice cream, and hamburgers. This English pianist could play all the popular dance songs of the day, and he played them without sheet music, strictly by ear. I was very impressed with his ability. Our social life at Clewiston was mainly centered around movies and weekend dances. If finances permitted, we would go to Palm Beach or Miami. There weren't all that many girls in Clewiston, so the competition was pretty severe. A weekend that stands out in memory was one involving a couple of buddies in Palm Beach. We were introduced to some lovely French girls from a wealthy family who had managed to escape the German occupation. The father of the girls was working in Washington, D.C., and his wife and daughters stayed in Florida during the winter months. We went swimming with the French girls during the early afternoon and then to a club where they were well known. We had refreshments and danced. The following day we met them again and took them to this same club only to find out that we had walked out on the check and our waiter had to pay it out of his own pocket—not a good way to impress the girls. We truly were embarrassed and had no intention of stiffing anyone, least of all the waiter. We dug deep in our pockets, paid the bill, and were more than generous in tipping our waiter of the night before.

We graduated in Course 12 on May 24, 1943. Getting their wings were seventy-two British out of the eighty-three who had started the course. Getting both Army Air Corps wings and RAF wings were sixteen of the seventeen Army Air Corps cadets. Following Wings Parade and the impressive Wings ceremony, the U.S. cadets were sent to Alpena, Michigan, for reassignment. I was with a group that went to Brownsville, Texas, where Braniff Airlines pilots trained us on DC-3s. We made several trips to the Panama Canal and to San Francisco via Los Angeles. It was at Los Angeles that one of our Course 12 classmates, Milt Steuer, was killed when his DC-3 crashed on a missed approach on a foggy night.

I completed training with Braniff and was sent to St. Joseph, Missouri, for instrument training. I was then assigned to the 3rd Ferry Group and checked out on the P-39, P-40, P-47, and P-51. I didn't check out on the P-38 because it was down for repairs. Most of my ferry flights were taking P-39s from Niagara Falls to Great Falls, Montana. I had a rather harrowing incident during that time. The route to Great Falls was over Fargo, North Dakota, just forty miles south of my home. On my first trip I was flying a P-39 with a two-hundred-gallon belly tank that made the Airacobra fly like a bathtub. I was able to give my parents a good show, flying low but very sedately because of the heavy-bellied tank. Later I was flying a P-39 with only a seventy-five-gallon belly tank, which made the plane much more maneuverable. I thought it would be a good idea to practice my slow rolls before I got to my home, so I tried one in level flight. Perfect. So I tried another one. Not so perfect. During the roll, the bulletproof glass behind my head came loose and punched a hole through the rear canopy. Whoosh! All of the airplane's delivery papers, my extra clothing, and my briefcase went through the hole with a giant sucking sound. To top it off, the plane stalled for some reason, even though I was well above stall speed. Perhaps my laundry was hung up on the tail. Who knows? At any event, it took two additional attempts to bring it out; by that time I was mighty close to the ground. I had no desire to follow through on buzzing the home farm.

I left the aircraft at Bismarck, North Dakota, where they had a spare canopy from a P-39 that had made a wheels-up landing and become an instant "spare parts" source. A couple of weeks later, I again picked up the plane and flew it to Great Falls. I never heard a peep from the brass about the incident and never learned how they replaced the missing papers. And I didn't ask any questions. My dad wasn't one to give up easily on losing something. He made inquiries in the area where my slow roll had taken place. Dad walked the banks of the Red River until he found my briefcase, a little worse for wear with a few scratches, but ready to go many more air miles. I have that briefcase to this day. We have traveled together for many years, flying aircraft to many places in the world.

Following the Niagara Falls duty I was sent to Reno, Nevada, for check-out on the C-46. After graduation I was given a ten-day delay en route, which I used to marry my college sweetheart. I then flew a C-46 along the southern route to Europe, down to Natal, Brazil, and then landed to refuel at Ascension Island, a "hunk of rock" in the Atlantic Ocean between South America and Africa; then on to Dakar, capital of Senegal, and, finally, to Marrakech, Morocco, in October, 1944.

During the next thirteen months I was based in Tripoli, Libya, and Casablanca. I checked out on the C-54 toward the end of the war and flew one back to the United States in time to celebrate Thanksgiving in 1945. I have a total of more than 1,700 hours of flying experience. I was discharged from military service in December, 1945, and returned to college a few weeks later to complete my engineering degree. I graduated in June, 1947 and worked for the same company in engineering management for almost thirty-seven years, until I retired at sixty-two.

William Slade

Altamonte Springs, Florida

Bill Slade was one of seventeen U.S. Air Force cadets at Riddle Field who trained with eighty-three British Royal Air Force cadets in Course 12, the first course with British and American cadets training together. He was assigned to the Air Transport Command of the U.S. Army Air Corps after he won both Air Corps and RAF pilot wings at Clewiston. Slade transported everything from C-47 Goony Birds to the hottest fighter aircraft of the era. Following air transport service in many parts of the world, he became a motion picture producer and director working primarily for the U.S. government in a civilian capacity.

When people hear that I spent time training in Florida with British RAF cadets, one of the frequently asked questions is, "How did you get along with the Brits?" The quick and accurate answer is "very well." Most of us in the small American contingent formed warm friendships with the British cadets. But after graduation they moved so far away that friendships in many cases faded away. Today I might recognize a British cadet's name but little more. There were a few times when tempers flared a bit because of some silly sort of thing. The British prepared food, pretty poor by our standards, but much to their liking, and when we complained about the taste we got a snappy reply: "We don't give fuck-all for your opinions."

I was one of the fortunate U.S. cadets to go to one of the British Flying Training Schools as part of the Lend-Lease Act. Those of us with prior flying experience were selected for a British school in Florida or some other spot in the States. We were virtually assured of not being among the washouts. "You are now *officially* under British command," we were told, although we continued to wear our Army Air Corps cadet uniforms. We had British officers and noncoms giving orders; our only American officer was a liaison officer.

British officers and civilians taught the ground school aspects of the program. For flight training our instructors were civilians and handled the Link trainer program (flight under instrument conditions). When we marched to classes, it was "by the left quick march" British style. We followed British style also in our inspections and physical training. There was a certain amount of rivalry between us and the British cadets and a few cases where we felt discrimination. By and large things went very well, much better than expected. The British were quick to make friends with us and with families in Clewiston and other communities. They were always being invited into homes for meals and social activities. The local girls found the Brits "romantic," with their unusual uniforms and British accents.

Dive-bombing with Coke bottles captured the imagination of the British fliers. They would carry a few of the curved green bottles when flying cross-country, tossing them at targets on the Florida landscape. Because of its configuration, the old 6 oz. Coca-Cola bottle made a satisfactory whistling sound as it was dropped over a herd of cattle. One great dive-bombing story involved a cadet who spotted an electric railroad handcar going down the track at a pretty good clip. This was a target that called out for all his skills. He made two very low passes at the handcar. Two men jumped off the car on the first pass. The cadet turned for a second pass and saw the railroad handcar men chasing it without gaining an inch. I have often wondered whatever happened to that little rail car.

Now I'll tell a dumb stunt story on myself. I must admit that in the exuberance of my youth I thought it was a good idea at the time. I was on a cross-country solo flight and was over the area of Palm Beach. Heading out over the Atlantic, I noticed a blimp (about the size of today's Goodyear blimp) flying submarine patrol. Hey, what a target for a hot pilot! I made a pass at the blimp and then decided it would be a dandy idea to add a little more color to the "attack." A loop around the blimp was just what the doctor ordered, and it was a good one. But had I miscalculated one little bit, there possibly could have been a loss of lives (including mine) and a lot of high-priced equipment wiped out. No way to win a war. Why didn't I think of that before? The answer was simple: We all pictured ourselves as hot pilots capable of taking on those Japanese Zeros or German Cross airplanes anytime or anywhere. So all we could do was fake attacks on targets of opportunity. Many times we "attacked" the B-17s that operated out of Avon Park, Florida. What if some of those gunners had cut loose with their .50 caliber guns with a few bursts? "Dear Mother of a Cadet: The United States Department of War regrets to inform you that your son was killed while pulling some idiotic trick in training. Should you have other sons, keep them away from matches, sharp knives, and airplanes."

There was one hotshot American trainee who had been a crop-duster in Texas and Oklahoma and thought he could fly anything made since the Wright brothers. Giving him flight training was like telling Babe Ruth how to hold the bat for a little more distance. This cocky cropduster, flying a Stearman biplane in primary training, was going for a beautiful three-point landing. He had to pull the stick carefully back to hold the plane just off the ground in the three-point position and came in somehow a little high. He didn't have the stick back all the way so he "horsed" it fully back. Had he been in the air, he would have executed a snap roll, maybe a good one. As it was, he got a snap roll all right but landed upside down.

The British taught flight training in Florida as though they were in England, simulating the wartime conditions over there. Our night flying was done under blackout conditions. No lights were permitted at the base—no floodlights, no runway lights and no landing lights used on the aircraft. Oil flare pots were lit—about six on each side of the runway—and we had to land by that amount of light only. Although approaching a completely dark ground and horizon with only those flare pots to guide you (they seemed like Zippo lighters), we got used to the British system and it worked out successfully. After training in the British low-light conditions, I would get confused when night landing at U.S. fields. The tremendous amount of lighting on the runways and other areas was unfamiliar and strange. Too many lights. Bats must feel that way, too.

At graduation there were eleven of the eighty-three British cadets who had washed out. All sixteen of the U.S. cadets did get dual wings, but in all fairness, they had considerably more flying experience than did the British cadets. A few of the British graduates would get pilot-officer commissioned status but most wound up as sergeant pilots. Class status apparently played a part in some respects, rather than pure ability.*

The avenue followed by U.S. cadets after graduation was generally the Air Transport Command of the U.S. Army Air Corps. We were sent to Billy Mitchell Field in Milwaukee, Wisconsin, for staging to the various air transport command squadrons. I went to United Airlines for transition to the C-47 at Denver. We had airline instructors and after graduation flew with the airline that had trained us. I flew with United Airlines, which hauled military freight across the country. Each of the commercial aircraft had one airline pilot and one military pilot and we split time fifty-fifty.

I was stationed in Chicago, flying round-trip from Chicago to Denver with a twenty-four-hour layover in Denver. Then, after two or three days, a round-trip to New York with a forty-eight-hour layover. What a tough life for a military pilot! But all good things come to an end and I was assigned to instrument school in St. Joseph, Missouri. I completed the course and started ferrying all types of military aircraft. First the C-47s and B-24 Liberators; then to

*Records show that fifteen of the seventy-two Royal Air Force cadets in Course 12 were awarded pilot-officer commissions. Fifty-seven graduated as pilot sergeants.

fighter transition school in Brownsville, Texas; and then back to Dallas to fly P-38 Lightnings, P-51 Mustangs, P-39 Airacorbras, P-47 Thunderbolts, P-63 King Cobras, and other aircraft. Eight months later I was picked as pilot on a C-47 group being transferred from Nashville to Great Britain. There were 150 crews with 100 new C-47s flying to England by way of Greenland and Iceland. We went over in October, 1944, on the last flight over the northern route before winter set in. Large reserve tanks of gasoline were placed in the fuselage, enabling us to fly that long haul from Goosebay, Labrador, to Reykjavik, Iceland. At one point in the flight, we received weather reports to expect hurricane force conditions. My own navigator told me we would make landfall in Scotland in forty-five minutes. We were still looking for land one hour and thirty minutes later. We learned that we were facing head winds of more than sixty miles an hour. Mayday! Mayday! Mayday! That chilling international signal of ships or aircraft in distress crackled over our radios. But the raging storm prevented ships from going out and nothing more was heard from the C-47, the third plane lost in our crossing.* Worn, weary, and reeling from the fierce North Atlantic operation, we arrived in England thankful that our losses were minimal— unacceptable to us, but minimal nevertheless.

There was no time lost before we were assigned to several bases in England and France. My group was sent to Cherbourg, France. We were stationed in what had been a prisoner-of-war camp for French soldiers held by German forces. We flew cargo back and forth between France and the UK, usually with gasoline and ammunition up to the front area and then returned with wounded to be treated or damaged war materiel to be repaired.†

We flew without radio because our transmissions could be intercepted and would lead German aircraft to our C-47s. Upward of 90

---

*One C-47 was lost, then rescued, on the Greenland ice cap and was later featured in a *Life* magazine spread. The crew had spelled out HELP in the ice cap snow. A second C-47 was lost flying from Iceland to Scotland, possibly because its crew had picked up a German broadcast beam designed to fool them into flying to a Scandinavian country. Cadets had been warned about these false signals.

†By the end of the war, the 302nd Transport Wing had evacuated more than 171,000 casualties by air.

percent of our flights across the English Channel were by instrument. On the rare clear day, we were constantly ducking other aircraft doing similar maneuvers. Better to be on instruments and not see a thing. As we said, we were better off flying "fat, dumb, and happy."

It didn't take ME-109s or FW-190s to knock out so many of our aircraft during the Battle of the Bulge. Heavy fog and generally inclement weather caused heavy losses on the ground. The fighter shortage grew so acute that transport command pilots with experience in fighters were sent to Scotland to ferry aircraft from Prestwick to forward bases in France. These were hazardous flights, as the whole of France and most of the United Kingdom were covered with thick layers of clouds at low levels and dense fog on the ground. There were no radio transmissions to lead you to your destination— you had to fly dead reckoning.

Fly or be court-martialed: this was the crystal-clear message I got at Prestwick when the base commander called out the pilot just in front of me and handed him aircraft papers and announced his destination. "When do you want me to go?" the pilot asked. "Now, for Christ's sake," the commander snapped. "You want to wait until May?" The pilot looked at the commander as though he were ready for a Section 8 [a discharge from service because of mental problems]. "But, sir, the runway is fogged in and there isn't fifty feet of visibility," the pilot said. "We can't fly in that." Without looking at the pilot, the commander turned to his operations officer and shouted, "Court-martial this man. Who's next?"

That was enough for the rest of us. We reluctantly took our papers, got into our planes, and made our way to the runway; we lined up and made an instrument takeoff. After about ten minutes on instruments we could fly to our dead reckoning destinations. Then it was a case of letting down through the overcast and hoping to find your base (am I in the right place?) and see to land. We solved part of this problem by flying in formation to our destination and taking turns letting down to see if there was room for the rest. We lost many aircraft in that blind hide-and-seek shelter operation, but we had to land. There wasn't enough fuel to return to England.

After the surrender of Germany I was sent to Oran, Algeria, on the Mediterranean to fly C-47s and C-46s, both cargo aircraft, across North Africa to Italy and then to Cairo. The aircraft would then be picked up and flown to the China, Burma, or India Theatres for war against Japan. I moved around a great deal: Oran, Casablanca, Tunis, Cairo, and Marrakech until the Japanese gave up and we were gradually sent home. I made the trip by Liberty ship.*

Liberty ship—you would not want to book a cruise on that ship. I boarded at Casablanca and had the good fortune to make it across. The ship would roll about 30 degrees either side of the vertical and then roll back to vertical. "Don't worry about that," a stocky sailor from Alabama reassured us. "When it rolls to 32 degrees, y'all got a problem. After 32 degrees it goes *all* the way over." The sailors were also helpful in telling us that Liberty ships were only made to last one voyage and the one we were on had made ten voyages. We only wished we could have gotten those guys in an airplane at least once.

One of the most unusual incidents I ever witnessed during my entire flying career was when I ferried a P-39 Airacobra from Great Falls, Montana, to Fairbanks, Alaska, to be turned over to the Russians. Over long flight distances, an extra fuel tank was fastened directly to the belly of the Airacobra. The fuel tank, known as the "bathtub" because of its shape, had no baffles to keep the heavy fuel from sloshing from one side to the other. Unless a pilot made a perfectly coordinated turn, he was in deep trouble. One ferry pilot in our group came into Fairbanks and made a bad turn, forcing the gasoline to push all its weight to one side. The P-39 snapped over on its back and the pilot, using his wits, applied power and kept the plane rolling until it reached an upright position and the wheels touched the ground. The pilot cut the power and rolled to the end of the runway, where he remained to calm his nerves and regain composure. Before the lucky pilot could taxi the plane from the runway, a cheering throng of Russian pilots ran up to the plane, pulled the American out, and carried him off on their shoulders. To them, it was the greatest flying trick of the century. I wonder how many Russian pilots tried that trick after that?

*A Liberty ship was one of over 2,700 merchant-class vessels built or adapted for transport of supplies and troops during World War II. They acquired the name "Liberty" in 1941 as a result of President Roosevelt's describing them as vessels that would "bring liberty to Europe."

Douglas Moore

College Station, Texas

"Doug" Moore was a U.S. cadet in Course 18 at Riddle Field, training with a large contingent of British cadets. It has never been clear to him (nor to many other Yank trainees) why the United States and Great Britain decided on the "mixed" courses in some cases and leaving the other courses exclusively British. Moore graduated with an engineering degree from the University of Tennessee in 1947 and earned his MBA at Wharton School of Finance, University of Pennsylvania.

I was about halfway through my Maxwell Field [Alabama] "Sam Browne belt" training when the captain of our regiment asked if any of us had a previous civilian flying experience. Noting that he said *a* flying experience, I answered "sure," acting as casual as Lindbergh. "OK," the captain said, "put your experiences down on paper and write an essay on why you should be selected for a special program." That was it. No further details were offered. I wrote a skimpy one-page outline about my flying experience (a joke), had a five-minute very military interview with the captain, and was excused. Ten days later I was told that I "had been selected." Selected for what? I thought. And to where? I was to be shipped out someplace on the Central Florida Railroad. So, with three years of engineering school behind me from the University of Tennessee plus a few hours of Civil Air Patrol, I arrived at Clewiston.

. I was assigned to Course 18, along with nineteen other U.S. cadets, mixed in with ninety British cadets of varying ages and backgrounds. My squadron included an older former policeman, a few cadets from upper-class universities (such as Cambridge), and all kinds of youngsters. I was the ripe old age of twenty-one and my Embry-Riddle flight instructor was twenty-nine.

Naturally, there was a great deal of curiosity over the British cadets. At the time the Course was formed in late 1943, Britain had seen more than four years of war and wartime rationing. Some of the British cadets who came to us were very pale and thin because of their improper diet. Their appearance in athletic clothes for sports activities was in shocking contrast to the Americans'. Their "shorts" were down to their knees. We were tanned, wore tennis-length shorts, and probably looked very flashy, or at least we thought so. But it didn't take long for that Florida sunshine and unlimited healthy food to bring the Royal Air Force cadets up to par and the change was like night and day.

After the first few weeks of training and a couple of weekends to look the girls over in Clewiston and Palm Beach, we came through to the Brits as being over-sexed, bragadocious, and sort of dangerous to be around, especially on leave. The Brits weren't unfriendly, but were feeling us out as well as sizing-up the towns in the area. The best way to really get to know each other was in sports. The Scots, rather short in stature, could run rings around us in soccer. Our best defense was to gang up with force, as in our football. The British cadets were top performers, not only in flying but in sports like water soccer.*

When we all took the bus back to the base from Clewiston, someone would start to sing and it was very noticeable that they were much better singers than we were—they could actually carry a tune. Unfortunately, some of the younger British cadets would come back from town liquored to the gills. Somehow they had not been trained in that aspect of social life. It seemed that we were more oriented toward group and team activities. They seemed geared toward individual efforts. This was significant in forming up bomber crews after graduation. [The success of several crew members combined in a large combat aircraft called for strong teamwork rather than egotistical, solo-pilot behavior.]

The British cadets looked forward to social events arranged by Clewiston residents at the Clewiston Inn and were drawn to what they perceived to be the "pub." One weekend I went with several of the British cadets to very hospitable events scheduled at Palm Beach homes of former Brits living in Florida. Dates were lined up to make us feel at home. However, a subtle distinction was evident: the university cadets (Oxford, Cambridge) went their own way, sort of

---

*Doug Moore's comments on the superior ability of British cadets in some sports was borne out by the performance of former Spitfire pilot Kenneth Edwards (Riddle Field Course 1). Edwards was a swimming force to be reckoned with, a true water polo champion for many years who always drew many U.S. and British fans when he put on dazzling displays at the Riddle Field pool. He was also outstanding at rugby. As a pilot he saw service in North Africa, Sicily, Egypt, India, and Burma.

smoothly avoiding their fellow cadets not born with the same advantages. Toward the end of our training there were ever-widening social gaps between the British upper classes and the working classes. (In fact, the Scottish corporal who ran their air equipment display was never taken into "their" culture. He was too much of a loner who really rolled his "r's.")

There was a running joke among the U.S. cadets who dealt with the Brits' seeming inability to cope with technical problems. What do you look for when something breaks down? If you are a Yank, you look for a pair of pliers and some wire. If you are a Brit, you look for a telephone. A British cadet could explain in detail how an airscrew (propeller) works in theory, or he could write a detailed paper on all aspects of flight. But when something broke down, the British flier's first impulse would be to call for maintenance. Then he'd stand back and watch what was going on. The American cadets, mostly Depression kids, would roll up their sleeves and take a crack at fixing whatever it was. Most of the time we could figure it out, take a few twists and turns, skin a few knuckles, and do a little cussin'. Chances are we had it licked.

At graduation time there were sixteen out of the twenty of us who finished. Fifteen of us were commissioned 2nd Lieutenants; one was made flight officer, not commissioned, but a grade above any enlisted rank. It seems that he had socialized too much with a girl in Moore Haven who ended up pregnant. Our U.S. adjutant told me that the cadet had "not conducted himself as an officer and a gentleman," and he was therefore denied officer and gentleman status. Reports came to me later that the flight officer died in a crash.

Graduation for the British cadets was not in many cases a happy time. Many who had their hearts set on officer status had to grit their teeth and check their emotions. When it all came out, the old "class thing" pushed aside grades and flying ability. Not always, of course, but more frequently than not the policemen and others of the working class had to settle for noncommissioned ratings. Out of the ninety RAF cadets that started in Course 18 there were eighty-two who graduated. Pilot officer status was conferred on fifteen of the British cadets, while sixty-seven were made sergeant pilots. I recall very vividly the anguish among the British cadets who claimed they were pushed out of a commission because they didn't wear "an old school tie" or their fathers didn't belong to the right clubs. This rift continues today among British graduates, according to people who have been in contact with them over the years. Very sad.

Following graduation I was then assigned, in typical Army style, to ferrying fighters from Great Falls, Montana, to Alaska, where they would be turned over to the Russians. But I was too tall (6 feet, 4 inches) to close the hatch so I was checked out on multiengine aircraft. I ferried new B-17s and B-24s to modification centers and then to the East Coast. Most of my time was in the Air Transport Command flying the South American route to Ascension Island, in the middle of the South Atlantic, and then to Accra in West Africa after refueling at Ascension.

As the war wound down I got a chance to be mustered out in early 1946 to continue my engineering program.

Fred R. Renshaw

Godfrey, Illinois

Fred Renshaw was a U.S. cadet at Riddle Field's Course 12, the first course that saw American and British cadets training together. He was a contributor to Roger Out, *a Course 12 publication prepared by a talented editorial committee of cadets (writers, cartoonists, and poets). This publication, dated May 21, 1943, was distributed at graduation time. The introduction read in part:*

> British and American cadets have worked together, flown and played together, shared the same rooms, eaten at the same table, argued, laughed at and with each other, and have found that they have a great deal in common.
>
> The rugged lips of the Northcountryman and the quicker lips of the Londoner now proffer strange oaths and expressions born certainly many wingspans from Bow Bells and Wigan Pier, while from American lips might be heard glibly falling, "Oh, good show, sir!"
>
> Our heartfelt gratitude goes out to our squadron commanders, flight commanders and (not least!) our flying instructors: also to those hard-working, long-suffering ground school instructors who struggled valiantly against the Sand Man on those hot afternoons, teaching us the mysteries of lift. Or was it "life?"

Being a cadet at Riddle Field was a lucky and happy experience for me, and the exposure to the British cadets was wonderful. I learned a lot from them and we got along very well together. Many of the British men I trained with were daring and fearless pilots, and I am sure they did their part in winning the war.

My wife was with me in Clewiston. We rented a house and I was with her every weekend, even though I had to stay in the barracks during weekdays. Often we would invite some of the British cadets for dinner. One of our little dinners stays in my mind. I had gone hunting in the swamps and bagged a few coots. They looked like ducks to me, so my wife roasted them for dinner. They were awful, but the British cadets ate them with enjoyment, feigned though it may have been, without a word of distaste. We've had many laughs over that "wonderful feast." Sometime later I learned that the coot is a fresh-water bird with unwebbed toes. The unwebbed toes should have been clue enough that I had not brought duck home for dinner!

As the oldest man in the American group, I was assigned to the Air Transport Command. I ferried A-24s, A-25s, P-47s, DC-3s, and B-25s. Later I was assigned to the North African Air Transport Command. Ferrying planes from the P-47 Republic factory in Evansville, Indiana, to Long Beach, California, resulted in my being chosen one of the old timers with his "head on straight," as we would say today—a "steady Eddie," as we said back then, a guy who could be trusted not to goof-off or screw up an assignment. My job was to impart dedication and attention to duty to the young fighter pilots who were ferrying the planes from Indiana to combat destinations. But the temperament of a fighter pilot is like that of the star college quarterback ("put me in, coach"), as he scrambles in the backfield and wins the game in the final seconds. You don't know what he'll do next. Bomber pilots are the defensive linemen, solid types who can hold a plane straight and level despite flak and enemy fighters until he hears that welcome yell, "Bombs away, let's get the hell outta here!" Thank our lucky stars for those wild and woolly, tough-to-tame fighter pilots, but they often wound up in Canada or Mexico because they didn't stick to their flight plan.

We always arrived at our destination each time, often with considerable anxiety. I would look back and see every plane perfectly in place as we flew along, like a group of little ducks following Mama. Wonderful. Then five minutes later I would be all alone on course and the young fighter pilots would be off on their own in those

P-47s. Farmhouses would be buzzed as a plane would fly at china-shattering speed ten feet above the kitchen roof of Mrs. Johnson's house outside of Phoenix as she was preparing dinner. The aggressive ferry pilots delighted in buzzing small airports, pretending they were strafing enemy ME-109s and FW-190s parked on the aprons.

Despite their "go to hell" attitude and wandering ways, these young pilots made their destination each time, pretty much on schedule, and safely. It all made me think of when I was in the Civil Pilot Training program (New York University), flying seaplanes off the Hackensack River, and attending ground school classes at NYU. My professor, Dr. Spaulding, would often say, "If you live past your very first solo in an airplane, you will live to be an old pilot." We would look puzzled at that bit of philosophy and someone would ask: "What do you mean by that, Dr. Spaulding?" Shrugging his shoulders, the professor would give a little nod and say, "Well, so many young pilots tend to buzz their girlfriend's home or maybe their own family home that they get a little too close." Wonderful words of advice that would probably be ignored by a fighter-pilot type.

J. T. Moore, Jr., M.D.

Algood, Tennessee

Dr. Moore is a retired physician who was a U.S. cadet in Course 17, formed October 4, 1943, and graduated April 15, 1944.

The British cadets had attended training school before coming to the United States. They went to ground school and had ten hours' flight time on the Tiger Moth before getting on a boat to complete their training in America. We U.S. cadets were handpicked because we all had at least forty hours with a private pilot license plus three years of college. It was apparent we were to "help teach them fly." It appeared that the early all-British Courses were comprised of what we would start calling the "elite," the Oxford-Cambridge upper-class types. As the courses rolled by, more and more working-class cadets appeared. One of my good friends was from Bristol and another from Edinburgh, both of whom I would call "middle types." It was widely believed that two Royal Air Force cadets in our Course were, as we said, of "royal blood" and they, of course, were given leadership levels.

The close proximity of Miami and Palm Beach had a great attraction to the upper-class RAF cadets and they made a big hit with American girls of wealthy families who wintered in the very social areas of Florida. When the RAF cadets checked into Florida training, they were rather gaunt kids, pale as bed sheets. In six months they had gained perhaps twenty pounds with sort of a Florida orange marmalade tinge.

We had two weeks off between primary and secondary and the RAF cadets were given a military air transport pass they could use to fly anywhere in the United States. Many of them really enjoyed these two weeks and saw a great deal of the country. After the war ended, both we and the Brits were busy getting adjusted to civilian life and I didn't follow up on my British friendships. One of my RAF friends lost his leg in action. Another close friend and I communicated for a short while until my last letter was returned with a note stating, "the establishment at this address has been torn down." My friend had given me the address of a pub where he was receiving his mail.

All in all, my experience with the British cadets was pleasant and memorable. They were admirable people with the courage that it takes to do what they had to do. Many of them had a "studied recklessness" in their flying habits—just the thing that makes great fighter pilots.

Harold (Hal) A. Jacobs

San Diego, California

After twenty-one years of active duty in the United States Air Force (originally U.S. Army Air Corps), Hal Jacobs retired with the rank of Lt. Colonel. He became a check pilot air carrier operations inspector for the Federal Aviation Authority. Over the years he checked out commercial airline pilots on the DC-6, DC-7, B-707, B-720 and DC-10 planes. Jacobs was one of seventeen U.S. cadets who trained in a course at 3BFTS, Miami, Oklahoma, in the northeast corner of the state, about twenty miles from Joplin, Missouri. There were about 110 British cadets in the course.

We got along "smashingly well" with the Brits. We all lived together and, of course, eyeballed each other a bit warily at first. Cadets, regardless of country of origin, tended to approach anything or anyone carefully and suspiciously. You learned to first get the lay of the land. Until then you were on your best behavior and proper decorum from A to Z.

To me there was nothing as important as completing the flight training satisfactorily and winning my wings. You know the old story: When a superior said *jump,* you were on your way up without even asking *how high?* So, in this best-behavior environment, we were very cordial, polite, and proper. The British cadets were the same, if not more so. The Brits were likeable blokes and we became very good friends with some of them, just as we did with our small Yank group. But it is true that on a Saturday night open post or at a Sunday afternoon movie in town, Yanks pretty much stuck together. I have thought about that in recent years when I visited the Brits who were in my class. Why didn't we hobnob during off-duty more than we did? Perhaps finances had something to do with it. We got $75 a month, three times what a British cadet was paid. We always had money for a hamburger and Coke (dry county and state), and the Brits didn't always have the money.

Oklahoma folks are very friendly to begin with and they really opened their homes to the British cadets. The first night at the field we were invited to a town dance, with parents escorting their daughters to the college gym. Matrons of the town would bring girls over to the cadets and introduce us so we could dance to the jukebox tunes—all very proper. The Oklahoma (and Missouri and Kansas) girls sure loved to be with the British boys. We Yanks were not such a novelty and we noticed the difference in appeal. Invitations would come into the base from folks in Joplin or Tulsa asking RAF cadets to parties as overnight or weekend guests. As I look back, it is most likely that those families who did the inviting didn't even *know* that there were American cadets on the base. When we got a seven-day break midway through the course, a few of the U.S. cadets would take a Brit home for a visit. I had so many plans for those few days at home that I never invited anyone along. Now I wish I had.

Sports brought the British and Americans together as much as the social activities did. We taught them basketball and softball. They tried to teach us soccer and cricket, with very little success—those games, so dear to their hearts, were simply not for us. A few of the Brits became quite good at basketball, although I must admit that we played a game something like ice hockey when it came to contact sports—Texas rules, you know.

As completion of the course neared, tension was high, with one exam after another. Ground school classes were tough. We had to learn all the British ways of navigation, gunnery, instruments, and everything else. With pressure mounting as the days raced by, many of us felt we were going to drop with exhaustion. We were so close to getting our wings and were constantly edgy—sort of like walking across a mine field, I guess. The big day arrived. With each of us having at least seventy hours of flying time before starting the course, all seventeen Yanks passed the course. Not a single washout. We were sad that about 30 percent of the British cadets washed out—they would disappear without a word. British policy was to get them on their way with all due speed. Then followed the Wings Parade, with a U.S. general making the presentations along with our RAF Wing Commander. American cadets were awarded both the U.S. silver wings [pinned left side] and the RAF wings [right side]. British cadets were awarded only RAF wings. The great day had passed and the tension drained from our bodies and minds—it was over. The Brits were gone for reassignment and we were off to get back with our Army Air Force, a separation I'll never forget.

Today I keep in touch by mail with about ten of my RAF classmates. Many of the British fliers were killed in action, an unbelievable number. Flying "goony birds" (C-47s), carrying troops and supplies, a large number were shot down like pigeons. During our flight training we flew in foursomes with one instructor. Two of us would be on the flight line and the other two would be in ground school. Then we would alternate. Two out of our four were killed in the big war. One died recently, leaving me the only survivor of our foursome. Our class suffered some of the heaviest losses due to more combat time than the others. Out of eighty-five RAF cadet graduates, only about twelve have been located.

I chose the offer to go to the British school for preflight, eager to become a red-hot fighter pilot and vanquish the Hun and all that stuff. We knew the British schools were all geared toward turning out pilots of single-engine aircraft. But a graduate of an Army Air Force school had a 50 percent chance of going to multiengine bombers or transports. My dreams of becoming that hot fighter pilot chasing ME-109s over Germany were just that—dreams.

We no sooner had our wings than we were all sent to the Air Transport Command ("Allergic to Combat"), which was a downer of the first order. As I cried, I had to admit to myself that I was lucky to have graduated with my wings and I should get on, as ordered by Uncle Sam. Today, alive and well, I feel blessed that things worked out as they did. I have had a flying career of twenty-one great years in the Air Force and eighteen years as a pilot on the big jets. Yes, everything did keep "coming up roses" for me. But it would have been great to have flown that P-51 Mustang just once.

∿ The Instructors ∿

It was an amazing concept, almost unthinkable and a complete turnaround. The United States Army Air Corps pilot training program was doing very well, thank you, in graduating about three hundred pilots a year at Randolph Field. The suggestion that raised a howl of horror was made in 1939: The Army Air Corps could deliver upward of *ten thousand pilots a year* (and ultimately more than 115,000 each year) by using civilian pilots as instructors in a beefed-up program. This "outlandish" idea came full blown from the fertile mind of General H. H. (Hap) Arnold himself, Chief of the Army Air Corps.

It would be completely in character for the perceptive Arnold to "hear" the roar of Luftwaffe engines getting closer and closer. One can imagine him using colorful language and pounding on the desk saying then and there that the Army Corps strength of 2,500 pilots was woefully short. What if America should once again be involved in a war of global proportions? It was impossible to think of such a thing, of course, said Arnold's critics. The Great War had ended just a bit more than two decades ago. And how could a bunch of civilian "crop dusters" be expected to instill the Army Air Corps tradition in carefully selected college graduates drawn to a romantic image created on the sound stages of Hollywood? The very thought of civilian instructors made hidebound Army Air Corps "lifers" shudder in their Sam Browne belts and double their rum-and-cola highball intake (a popular drink in that era) at the various officer clubs.

The civilian instructor idea drew little applause. Consensus among career Army Air Corps officers stuck like a needle on a broken record: It won't work. It won't work. Building Air Corps strength could succeed only by putting cadets through training conducted by Air Corps pilots who themselves had been trained by Air Corps pilots. Anything else was a foolish dream. "Damn it," said a Maxwell Field major, "this civilian instructor crap is like expecting a kid on a bicycle to teach people to drive cars. And they're gonna learn that the whole friggin' idea will be kicked out before it gets started."

Then, even worse, Arnold's pipe dream went on to claim that not only would civilian pilots train Army fliers, but civil air schools could design the entire program for young cadets. Army brass would, of course, handle service matters and discipline but the total plan would be a joint army/civilian air defense endeavor. Despite strong opposition and sharp criticism, the plan went forward, limited at first to civilian pilots teaching primary flight training to Army Air Corps cadets and then expanding that teaching to basic and advanced. Some civilian pilots even taught cadets to fly fighters and bombers in the Army Operational Training Units.

Colonel Robert L. Scott, A.C., a veteran pilot who learned to fly in civilian life, had high praise for civil air involvement in the Army Air Corps training effort. He wrote in 1944:

> Whenever you read that another thousand-plane raid has blasted the hell out of Germany and that another Nazi city is missing or that a long-distance raid has burned an Axis oil field like Ploesti, be sure to remember that most of those planes were piloted by some kid who learned to fly in a *civil school*.
>
> His civilian instructor was one of those "puddle jumper" pilots who simply had the love of flying in his blood and the opportunity to fly nothing more than a "hangar door with an ice cream freezer power plant."
>
> Those "puddle jumper" pilots and all other civilian instructors have delivered when the chips were down. They may not ever have seen an enemy, but they have aided just as much in shooting down enemy planes and blasting cities as the men *they trained* to fly the fighters and bombers.

Embry-Riddle's BFTS instructors at Riddle Field and Carlstrom Field, all civilians, were what Colonel Scott wrote about. They came from all sections of the country to Arcadia and, later, to Clewiston. They headed to the Florida training fields to form John Paul Riddle's team of assorted pilots with their wide range of backgrounds and flying experiences. Some left their homes because flying was their calling and central Florida was one place to meet that itching need to "fly your airplane and unveil the true face of the earth."* Oh, sure, there were those persons of a poetic bent who were drawn to the controls of an airplane in part by the captivating words of the French flier-writer-philosopher whose best-selling books glorified the freedom of flight. But most of the instructors were typical of men in any occupation during those times. They wanted to instruct because that's what they could do best, and they had to make a living. They were not in the least romantic, but clearly realistic.

"Look at the addresses on this instructor roster sheet," a Riddle Field clerk said. "My god, these guys come from all over." Well, not quite all over, but a reasonable amount of "all over." A sampling shows Greenwich, Connecticut (Amoss); Sarasota, Florida (Ahern); Kenmore, New York (Bennett); Woonsocket, Rhode Island (Bishop); Perry, Michigan (Bridger); Brooklyn, New York (Brittain); Spring Valley, New York (Carlson); Washington, D.C. (Cockrill); Bloomington, Illinois (Davis); Peru, Indiana (Fair); Leominster, Massachusetts (Flynn); Montpelier, Vermont (Garcia); Elmira, New York (Graves); Lynchburgh, Virginia (Hawkins); Clanton, Alabama (Hayes); Atlanta, Georgia (Heffner); South Gate, California (Hunziker); Ashtabula, Ohio (Krell); San José, Costa Rica (Lyons); Penn Run, Pennsylvania (Moore); Tupelo, Mississippi (Morrisson); Cincinnati, Ohio (Mougey); Vineland, New Jersey (Piermattei); Dubuque, Iowa (two Richardsons); Camden, Arkansas (Sanders); Culver City, California (Smith); Nashville, Tennessee (Veltri); Youngstown, Ohio (Walsh); and Greenwood, South Carolina (Witt).

And so many more instructors were on those lists with different personalities, from what former RAF cadets report, but all had the desire to bring "their students" all the way through to Wings Parade. This was an impressive number of instructors at first glance, but operators of flying schools who received contracts to train pilots for the military soon found that the nation's pool of instructors was too shallow to supply the needs of an Army Air Corps caught up in explosive growth. John Cockrill, the first instructor at Carlstrom, looked back on the severe shortage of instructors when he spoke in the mid-1960s to Tony Linfield, a Course 18 RAF cadet who had been named best ground school student in his Course. Linfield had traveled from

*Antoine de Saint-Exupéry, *Wind, Sand and Stars*, 1939.

England to renew friendships that he had made twenty-five years earlier. Sitting in a restaurant-bar in Clewiston, cowboy boots resting on a table top, Cockrill reflected on the problem that developed toward the end of 1941: securing and retaining civilian instructors for the schools. With eyes half closed on a humid afternoon, Cockrill could just be heard above the whirring of overhead fans:

> You see, before the war, before Pearl Harbor, the people in this country just sort of sat on their asses and did nothing. Then all at once we had all of those big ships shot up.
>
> We lost most of our navy and a lot of our aircraft and we didn't have the set-up to rebuild real fast. We didn't have many flying instructors either. Now when I got my pilot's license in 1937, my number was 48,487, which means that only 48,486 people in the country had been licensed ahead of me. And that includes the Wright brothers.
>
> So I don't imagine that at the time of Pearl Harbor there were more than 5,000 active civilian pilots in the United States, and many of them were pretty old. Then the war comes along and we don't have any pilots, even though we did start to be aware of the shortage in 1939.
>
> But there was the college program in place to train pilots and the military was snapping them up. We had a real problem, so we started training our own instructors and at any given time at Riddle we had twenty to thirty people in the flight instructor school. The incentive was there, the money was good, so if anyone wanted to fly we'd get an instructor to solo him and then we'd enroll him in our instructor school.
>
> The Air Corps allowed us to give a student up to forty hours instruction under good teachers and if he couldn't make it we'd wash him out and then re-enlist him and give him forty *more* hours of instruction. That's the only way we could build up our instructor base. Before we graduated our own instructors, we couldn't retain enough of them to keep the school going, particularly when the British started coming over for training.
>
> Eastern Air Lines, National, and other airlines were picking away trying to grab instructors that they needed. They would have their scouts drifting around town, meeting instructors, and signing them up. So we made a rule. If anyone wanted to leave, he had to bring in a replacement for himself.
>
> Of course we had our own recruiting program going. We would try to get instructors from other flying schools and some even from a glider-training school. These were not easy times, but they worked

out. By the end of the war, the country had more instructors than it had had pilots at the start of the war.

The Air Corps tried to send me to balloon school. But I didn't want to learn anything about balloons. I wanted to fly airplanes.

Although Riddle could not properly be classified as an instructor at either Carlstrom Field or Riddle Field, two of his flying schools during World War II, his influence, direction, and leadership of hundreds of instructors and cadets—American and British—call for his listing in this section. This places him where he belongs, among the men he always felt he understood and knew best: pilots who flew the U.S. mail or barnstormed around the country in the 1920s and '30s; pilots who taught others to fly in small airports in Kansas, Louisiana, Texas, or Montana; the stunt pilots who doubled for Buddy Rogers, Richard Arlen, and Gary Cooper in the Howard Hughes classic *Wings,* or those civilian pilots who could only build up time on weekends, away from a regular workplace.

John Paul Riddle was born in 1902 and died April 6, 1989. As his life placed him among those with whom he had fraternal bonds— his instructors—his death placed him with twenty-three young Royal Air Force cadets who lie with honor at Arcadia's Oak Ridge Cemetery, men who died while training at Riddle or Carlstrom Fields during World War II. The headstone of John Paul Riddle is there with those who followed his own calling, a love of flying and love of country.

Few people have followed the career of John Paul Riddle any closer than Howard Melton, Arcadia historian and collector of Carlstrom Field memorabilia. Days after Riddle's death, Melton wrote this story for the *DeSoto County Times:*

John Paul Riddle was a pioneer in the field of aviation history. He declined appointments to West Point and Annapolis, beginning his long trek to home in the sky at age 18, just 17 years after the Wright brothers made their initial flight of 120 feet in 12 seconds.

His first opportunity came when he was sent by the Air Corps to mechanics school at Kelly Field, San Antonio, Texas. After completing mechanics school, Riddle went to Chanute Field, Rantoul, Illinois, where he was assigned to flying school at Carlstrom Field.

At Carlstrom, Riddle piled up more flying hours than did other cadets because he took hours that other cadets passed up in favor of free time to enjoy themselves.

Following Carlstrom, Riddle went on to Post Field, Lawton, Oklahoma, where he learned the value of sound aircraft construction.

Melton's story goes on to tell about Riddle's career after his two-year Air Corps service. He began barnstorming with a $250 Jenny in a dozen states, giving rides to passengers for a few dollars ("Sometimes pretty girls got to ride free," he said). After some years of engaging in various aviation-based enterprises, he and one of his former students, Higbee Embry, formed the Embry-Riddle Company to sell aircraft in Ohio, Indiana, and Kentucky. In October, 1939, Riddle formed a partnership with John G. McKay, a prominent Florida lawyer. The Embry-Riddle School of Aviation was born.

Len Povey and H. Roscoe Brinton are two other men who attained legendary status in the 1920s and '30s. Pilots of those early days gathered to do a bit of "hangar flying" and recount stories of their exploits. They were two cigar-smoking, bourbon-sipping Embry-Riddle veterans who "flew anything with a wing and an engine." Povey and Brinton were brought into the organization not only for their flying abilities, but also for their skills in getting superior performances from instructors.

John Paul Riddle, in 1940, appointed Leonard J. Povey as director of flying and Roscoe Brinton to Carlstrom Field as assistant director of flying. Brinton had won many commendations as an instructor for Embry-Riddle at Miami Municipal Airport. He later became general manager at Carlstrom. At the close of Povey's career with the flying school, he was vice president in charge of flying operations. Soon after Povey was hired, he reached out for experienced instructors Willis Tyson, Jack Hunt, Tom Gates, Joe Horton, and Wyman Ellis. The flying school to train military pilots was on its way.

Brinton's early career could have furnished inspiration for dozens of Warner Brothers movie scripts. He and his Brinton-Bayless Flying Circus headlined acts all over the country during the heyday of air races throughout major cities. Speed champions such as Roscoe Turner and Jimmy Doolittle awed crowds of thousands as they roared their pocket-size airplanes around shivering pylons. Then Brinton and his "death-defying, fearless fliers" would put on their wing-walking, low-flying, delayed parachute jumps as spectators screamed in fear for the lives of these "utterly reckless" stunt men and women. Of course, the aerobatics were so well orchestrated and rehearsed that accidents were almost unheard of and these flying circuses were at the height of their popularity. (Brinton's son, H. Roscoe Brinton Jr., was a flight commander at Riddle Field.)

Len Povey, like Brinton, was one of the great barnstormers of his day. He signed up with Brinton's flying circus and then, in 1933, was asked by the Cuban government to reorganize its air force of 22 planes. Povey spent four years teaching Cuban pilots to fly and built up a sizable air force in those pre-Castro years. It was in Cuba that Povey bailed out of an airplane for the second time in his career (the first time, he said, was "done in the spirit of self-preservation"). He had his plane in a dive over Havana when it came in contact with a large buzzard, which caused the left aileron to be sheared off. Povey got the plane on its side and managed to get it out over the water before he bailed out. He was decorated for getting the plane away from the city. When the Red Cross official gave him the citation, Povey said, "You know, I must tell you that when I bailed out, I had no idea where that plane might crash." The Red Cross representative said, "That's all right. We'll make your award for honesty instead of heroism."

Povey stayed in Cuba until 1938 and won the Mexican trophy twice at the Miami Air Races. He remained an honorary captain of the Cuban Air Force while he was at Embry-Riddle, and went back to Havana once each year as technical advisor.

"Slow Walker" was the nickname for the twenty-seven-year-old advanced training instructor with the soft Florida drawl. He had a closer affinity with the British cadets than did other instructors. The heritage of John Seymour Weston Davis was English through and through, as English as cricket, a pint of bitters, and Charing Cross. Davis's mother and father came from England to the United States and settled in Fort Myers, where John was born. The family later moved to Okeechobee, at the north end of Lake Okeechobee. His mother retained her British citizenship throughout her life.

How he became known as "Slow Walker" was similar to tagging a person six foot seven inches tall with "Shorty" or a two-hundred-

and-fifty-pound heavyweight "Slim." When Walker was a youth, he worked for one of the railroads for a short time, and soon was ribbed by older workers because he moved about so quickly on the job. One of the railroad workers noted from Davis's time card that the initials S. W. fell between John and Davis. "Look here," he said to Davis, "you're not living up to your name. That S and W is supposed to stand for Slow Walker and you move around faster than a jackrabbit." The nickname has stuck to Davis for more than sixty years. RAF cadets quickly picked it up and use it as a salutation whenever the Davises make one of their frequent trips to visit old friends in England.

Davis was among the first instructors at Riddle Field when Courses 1 and 2 were being held at Carlstrom, waiting completion of the field. He was an advanced instructor from 1941, assigned to teaching odd-numbered Courses 1 through 19. (At the start of training in 1941, instructors were given five students per Course; that figure dropped to four students later on.) When Davis signed on at Riddle, he brought with him almost six years of solid flying experience. Charles Miller had soloed him in 1936 at a small airport in Okeechobee, Florida, and the two young men, Davis, twenty-two, and Miller, twenty-nine, barnstormed throughout Kentucky, Tennessee, and Missouri.

"We would fly around some small town to get attention from the folks below and we'd land just about any place we could," Davis said. "We would put up a sign offering flights around the town for whatever the traffic could bear. Charley's wife would sell the tickets and Charley and I would take turns flying the plane. I must have flown about a thousand passengers before I ever got my license."

After a few months as an advanced instructor at Riddle Field, Davis was made assistant flight commander with Sam L. Schneider. The flight commander was Davis's old barnstorming partner, Charles Miller, who had taught Davis to fly. Instructors in the advanced flight were Jean Reahard, R. A. Westmoreland, Noel Ellis, Keene Langhorne, Charles Bing, Henry J. Middleton, Lou Place, A. F. McGravey, Donald C. Day, and Robert V. Walker.

By 1943, Davis was promoted to flight commander when Miller moved up to advanced squadron commander. "We always got a little boost when those promotions came along," Davis said, "maybe twenty dollars a month or so. As I remember, we started out at about $300 a month and it ranged up to about $500 a month, depending on promotions and the length of time you had on the job. I didn't know at the time that there was that much money in the world, because I remember when five bucks a week was pretty good." Recalling some of the Riddle Field days, Davis said he could remember very few students washing out in advanced training. "Most of the boys who did wash out must have been in primary or basic. When they got up to me in advanced, it was pretty likely that they were going to go all the way." Davis is remembered as being an understanding, patient teacher by many former cadets and fellow instructors. "Johnny Davis was a man you had to admire," said Phil Kinsey, who was on the Riddle staff with Davis. "He knew how to get his point across and he was down to earth. The boys really got along fine with him."

Course 3 cadet Bob Richardson came from his Glasgow home in 1993 to attend a reunion of 5BFTS Association members at Clewiston and was surprised when Davis picked him out of the crowd. Richardson, an early Davis "fan club" member, said: "I was standing in the shade of a large tree when I was suddenly embraced in an absolute 'bear hug.' A voice said, 'I just have to get hold of the man who writes those marvelous letters to me.' I turned around and there was Johnny 'Slow Walker' Davis, looking so much younger than his years. It seems to me that all our former instructors look younger than the average person in his seventies or eighties."

Freddie Stewart, Course 1 cadet, had high praise for Davis: "He was a steadying influence on all of us; we were the first group of lads assigned to the 5BFTS and needed all the encouragement we could get. Johnny Davis was first-rate when it came to putting us at ease." Davis obviously has a knack for putting people at ease. This is confirmed by a first-hand report from Mrs. Davis: "Johnny even taught me to fly. He gave me five hours of instruction back in 1946 and then he told me to take it up alone. I must have done all right, because when I started to land, he waved me off and I went around several more times."

When Course 19 graduated on August 26, 1944, with seventy-seven of the original ninety winning RAF wings, Davis was recruited by Eastern Air Lines. He spent thirty years as a top pilot with the Miami-based carrier, retiring at sixty, the mandatory retirement age for pilots. Davis then went back to one of his earlier pursuits, raising cattle for the beef market on his 1,000-acre ranch near Clewiston.

"Here's a little question for you," Davis asked his interviewer. "How many pounds of grain does it take to make a pound of beef?"

"Oh, say, about maybe a couple of pounds?"

Davis pounced on that answer. "You fail," he said. "It takes eight pounds."

There's always a teacher in every crowd.

When he is not too busy, Davis still uses the runway at his ranch to enjoy his two aircraft, a Mooney and a Piper Cub. Mrs. Davis sums it all up: "He's always known how to fly just about everything at any time."

When it comes to the early days at Carlstrom Field, you also have to turn to the name of George Ola. He was there at its rebirth in early 1941, a new edition of its old glory days of the Great War. At twenty-seven the erect, movie-actor-handsome, Geistown, Pennsylvania native became commander of cadets at Carlstrom Field. Second Lieutenant George Ola was known as one of the United States Army Air Corps *Wunderkinder.* He joined the Air Corps as a private in 1934 after two years at the University of Pittsburgh, Johnstown campus, and served more than two years in Panama as an assistant crew chief. In 1938 he was accepted as a flying cadet at Randolph Field, Texas, where he did his primary and basic training. He then completed advanced training at Kelly Field, Texas, winning his wings and attracting favorable attention from Air Corps superiors. Now commissioned, Ola gained his instructor experience at Randolph Field, where he trained cadets in the basic phase of flight training.

The slim, reserved young lieutenant was a popular figure at Carlstrom Field, making a near-meteoric rise in both responsibilities and promotions. He went from second lieutenant to major, a three-grade jump, in less than a year, and was made commander of the field in 1942. Ola had a determination that may have stemmed from his boyhood days in the Johnstown area, where survival has been a watchword passed on to generations of youngsters since 1889, when one of the nation's greatest disasters, the Johnstown Flood, claimed 2,200 lives. Any boy raised in the coal-producing, steel-making region was no stranger to calloused hands and more often than not was an eager participant in the brawls that erupted when youths from rival towns "had at it," as the Irish folk of East Conenmaugh and Nanty Glo would say.

Ola always radiated confidence in himself. One primary instructor remarked, "George Ola might appear a bit cocky to some people, but, no, you just gotta figure that the guy is only self-assured to beat hell." Whipping an aircraft around in eye-popping aerobatic maneuvers on the days cadets were getting final check rides, Ola would draw applause from envious cadets and smiles of admiration from seasoned pilots. His skills were long remembered after cadets left Carlstrom.

Former RAF Warrant Officer John C. Edmunds wrote from London:

> Captain Ola always gave such a dazzling display of aerobatics to cadets going through final check. I have a sneaking feeling that he was responsible for my desiring to become a Spitfire pilot, a desire that was realized.

National tribute was paid to the flying skills of Captain Ola in a 1942 edition of *Collier's* magazine in an article by Corey Ford and Alistair McBain, "Take it, Mister." The article featured Carlstrom Field. The authors described an illustrated handbook on the various Ola maneuvers, but the whimsical writing team pointed out jocularly that "there is no mention of a Whifferdill." A Whifferdill, they go on to explain, is something that no one else but Ola would want to do:

> You start a loop and as you complete it, you roll over on your back and come through another loop underneath, then do an Immelmann, and end up going where you started, only in the opposite direction.

The writers asked, "Is the Whifferdill possible? Perhaps it is and maybe it was a Whifferdill that Captain Ola was doing over Carlstrom Field that night in the deserted twilight sky."

> The sun was down and long blue shadows lay over the cadet barracks and palms and empty lawns; but the last searching rays of the sun still played on the solitary training ship high above us.
> We could hear the drone of its motor, increasing and fading and

increasing again. The ship rolled and plunged luxuriously like a dolphin playing in a phosphorescent sea. As it joyfully played, there was activity below. The sidewalks and verandahs and lawns in front of the barracks began to fill with awed and silent cadets, emerging from their rooms with books in hand, or hurrying dripping from their showers, or piling out of their bunks in undershirts and shorts.

They stood in the dusk with upturned faces, staring at the sky. Tennis players halted their game, rackets still poised. Six cadets walking off gigs (marching punishment for minor infractions) kept their eyes aloft as they marched up and down.

What is a Whifferdill? Only Captain Ola knows for sure.

Ola was transferred from Carlstrom Field with well-deserved recognition from the Air Corps Command, but not before he married Ruth Pemberton, a former Miss Arcadia. A fellow officer remarked, "Hey, George got credit for a great job at Carlstrom, but that was only the second best thing he did. The first best thing was getting Ruth to marry him."

After several stateside assignments, Ola went on combat duty during the Korean War, flying combat missions over enemy territory, and was confirmed for shooting down an enemy aircraft over North Korea. His decorations include the Distinguished Flying Cross, the Bronze Star, the Air Medal with clusters, and numerous other awards. After several other posts in the United States, Ola was sent to England to be deputy wing commander of a fighter wing for three years before returning in 1960 for retirement as lieutenant colonel. At eighty he was still flying small aircraft from a landing strip at his home in Arcadia, back where the most important phase of his long career really got started.

What did it take to be an instructor during the wartime years at Clewiston? Phil Kinsey, Fort Myers, Florida, a former Riddle instructor and native Floridian of long family tradition in the state, gives his views:

To start with, you had to remember that you can learn to fly, but teaching it is like an art. I figure that if you take a dozen or so pretty good pilots, you'll find a couple of them outstanding as instructors and the rest ranging anywhere down the line.

I don't know the answer as to what goes into what you must have to be a flight instructor. One thing I know—you don't have to be of the so-called hot pilot category. What it takes is solid experience as a pilot, depending on what level you are teaching, and patience to deal with the mistakes that the young kids are going to make.

I always avoided loading down the student pilot with too many things at the same time and bringing too much pressure on him. You had to figure if a kid kept doing something wrong, time and time again, maybe you ought to look at what *you* were doing. Was I getting my point across? Maybe I ought to take another approach.

The Kinsey system followed quite closely the ideal textbook drill, but there really were no norms when it came to style. What worked for one instructor would fizzle for another. One Carlstrom Field instructor took a hard-nosed approach with all his students right from the start, the minute he shook hands with the new cadets. Was he popular with many of the boys (first RAF and then Army Air Corps) that he trained over a two-year period? "Hell no. I was about as popular as a shut-off man from the gas company. But I'll tell you they were damned glad that I wasn't their buddy when they went on to basic. I wish I had saved some of the letters I got from my students after they went into combat. You should read how they gave me credit for hammering into them the real fundamentals of flying. You know, the kind of stuff that could save a guy's ass when he was on the spot." Then there was Ralph Cuthbertson of Carlstrom, who stressed the "importance of mental conditioning" in taking flying instructions. Today, Cuthbertson could win instant fame, fortune, and television renown as an author of "you can do it if you try" books. He set forward his "mind motivation" principles each time he was assigned five students (the standard number) with each new class:

Failure to instruct properly is all in the mind. All in the mind. That would be my problem. Failure to learn is all in the mind. All in the mind. That would be your problem. You will be here in primary for but a short while, and then you will go on to basic at another field. Tell yourself that you *will* go on from here. You will *not* wash out. Remember that it is *all* in the mind.

I will tell you myself that my teaching will *help* you learn and I must remember that I *cannot* fail my responsibilities. It *is* all in the mind.

Cuthbertson, who later shifted to Riddle Field when RAF training closed down at Carlstrom, packed up his psychology 101 course syllabus and took it along. It worked, too. He had one of the lowest "elimination" (washout) rates among more than one hundred instructors. Former Carlstrom cadet David Ferguson of Dublin, Ireland, recalls the time he spent with Cuthbertson. "He was one of those persevering types who always seemed to get all five of his primary students through and into basic. It must have had something to do with his 'all in the mind' lectures." Yes, perhaps. Then it could be that Ralph Cuthbertson was one of the finest pilots and most convincing mentors who ever signed on as an instructor. All in the mind? Oh, sure.

Howard L. Graves, Bradenton, Florida had one of the longest service records among the instructors: May, 1941 to October, 1945. He taught primary at Carlstrom Field until November, 1943, transferred to Riddle Field, taught primary and advanced (combined with basic), and became a primary flight commander. Graves soloed at sixteen and had his private pilot license by age seventeen. He went to work for Piper Aircraft Corporation, Elmira, New York, in 1937, after graduating from Elmira Academy and Elmira Aviation School. As Graves described his career,

I was responsible for a complete check on every Piper aircraft that went out the door to the flight test area. Joe Horton was a resident Federal Aviation Agency inspector who made spot checks of our Pipers. He left Lock Haven to go down to Carlstrom Field to head up inspection of the Stearman PT-17s, and finally convinced me to go down to Arcadia and work with him.

Joe kept telling me that Carlstrom was going to be the West Point of the Air, a permanent flying school, and you really had to believe him when you saw the first-class layout—the circle of white concrete barracks, the tennis courts, and the beautiful swimming pool. Someone had planned a flying school that would last for many years.

I worked for about six months as a maintenance inspector and rode on the test flights with Clem Whitenback, a well-known aerobatic pilot before coming to Carlstrom. Clem suggested that I take the army refresher school course and that was all I needed to hear.

I went through the thirty-hour Army Instructor School at Carlstrom and taught primary, logging more than 1,900 hours of Stearman instruction. What I really wanted to fly was the AT-6, a beautiful advanced trainer. I got that chance when Len Povey, director of flying, let me transfer in November, 1943 to Riddle Field.

The last assignment that Graves had at Carlstrom was in Class 43-C, Flight 5. His fellow instructors were John S. Ayala, flight commander; 'Robert Greer, assistant flight commander; Robert Cross, E. T. Glenn, Thomas Hellender, O. L. Hutchins, Herbert Lindsay, Eugene Poymter, Thomas Taylor, William Tanner, O. E. Van Schaich, and Martin Walsh.

It was interesting to note the difference in the U.S. Army curriculum (Carlstrom) and that of the British RAF at Riddle. The British felt that timing maneuvers such as the two-turn spin and the snap roll were of little value, while the Army Air Corps required them.

A compulsory RAF training maneuver was operational low flying. The British wanted full power flight below the tops of the palm trees. An instructor would close the throttle and declare a simulated forced landing; the cadet would immediately gain as much altitude as possible with excess air speed. He would open the canopy and establish a glide to an open area. The open canopy would, of course, help prevent entrapment if a nose-over occurred in the actual landing.

Another maneuver was the precautionary landing, which resulted in some serious accidents. The object was to make the shortest possible landing over a canal bank on a practice field. A very slow nose-up full flap approach could be made, controlled by a little power. Heavy braking was used with caution against nosing-up. The RAF advanced AT-6 program created some very good pilots in a short period of time.

Instructor Bill McGallard had an interesting day when he took up a cadet for instruction of slow rolls. At altitude, Bill asked the cadet to execute a roll. When the plane reached the inverted position, the nose dropped. "Keep the nose up," Bill yelled. Bill gave assistance and rolled the plane back to level. Then he noticed that he was flying alone. No one was in the back cockpit. He made a quick bank and peered down below. There was a snowy white parachute slowly floating toward earth. The cadet later explained that he had forgotten to secure his safety belt. He probably never did *that* again.

I think that we all know how reckless we can be when we are young. I was in my early twenties at Riddle and really enjoyed the aerobatics with my cadet students. This was not true of some of the older instructors. If I had a fast learner cadet, I would show him inverted spins, inverted snap rolls, and snaps on top of a loop, among other aerobatics.

Phil McCracken, a young flight instructor, and I would stay after the cadets had finished for the day and give each other some hood time.* The Stearman had a canvas hood that pulled from behind the seat of the rear cockpit. You pulled it over your head and latched it above the instrument panel. The panel was all you could see.

I was in the front seat and Phil was under the hood. We had finished a half-hour hood time for Phil when we discovered that neither of us had brought a car to the field and the last bus left for Clewiston in about a half hour. We didn't have time to land and change seats so I could get in my hood time and I didn't want to lose out. There was only one answer: we would switch seats in the air. In that way I would get in my time and we could get back to the field before the bus left.

I closed the throttle and trimmed the airplane for a 90 mph glide. Using the wing hand grips I struggled out on the left wing, locking my left arm around the center section strut. I could then reach in the cockpit to the control stick.

In the meantime, Phil got on the right wing and then into the cockpit up front. Somehow I managed to move from the left wing to the rear cockpit. What a crazy stunt. I shudder every time I think what could have happened.

When the war ended and the field closed down, my wife and I got into our Ryan airplane and flew back to Piper Aircraft, where I worked as flight operations manager until the last day of 1979, when I retired. Soon after we moved to Florida. Those friendly "crackers" told us that once we "got sand in our shoes" we would always come back. They were right.

Graves received a letter in 1993 from one of his former RAF students, who expressed gratitude for his training. Former cadet David Morgan, Riddle Course 22, wrote in part:

I do not suppose that you remember me as an individual after all these years, but, as one of your ex-pupils, I am glad of the opportunity to tell you that you have certainly not been forgotten, and that those days in Clewiston are still remembered with nostalgia and affection.

*Hood time was used to simulate instrument flying conditions. The person under the hood would be flying only by instruments.

You can't imagine how important was your quiet, steadying influence, because my father had died and the news had been kept from me and from my mother. Unfortunately, a friend wrote a note of sympathy to me, which came as a tremendous shock when I opened the letter.

Your last flight with me was on 21 March 1945, just before my final check by Flight Lieutenant Stephen Harvey, RAF. My log lists that we spent two hours in an AT-6 doing stalls, steep turns, low flying, precautionary landings, and forced landings.

I imagine I was sweating a lot after that, but you must have licked me into shape, because I got my wings the next day. Now is my chance to say thank you for the part you played in getting me that coveted emblem.

If you had to pick a prototype for an ideal instructor, you wouldn't go wrong in selecting Jim Cousins, one of the most popular, admired, and respected pilots at Riddle Field. He later became an Eastern Air Lines captain and retired in 1977 after thirty-three years with the airline, having logged so many flying hours that the adding machine ran out of tape.

Jim Cousins's long flying career began in 1937 when he was a teenager on Florida's west coast. He was the first person to solo at the little Venice Airport, taking off with a Piper Cub when he was eighteen, fully determined that he had found his niche in life. It wasn't long before Cousins and some friends turned their attention to a location near the Sarasota-Manatee County border, where there had once been a mid–1930s type of airport. Located on the site was a beer tavern called the Mystery Ship, because it featured a World War I relic on the roof, an old flying boat with a wing span of about sixty feet. "When we got there, there was a jalopy track around the old field that was covered with ruts caused by hundreds of tire tracks," Cousins said. "We did some flying out of there for a few months and then I went over to Miami in 1940 and hooked up as an instructor with John Paul Riddle."

With the opening of Riddle Field and the start of the British RAF training in 1941, Cousins moved to Clewiston, serving as instructor on Courses 1, 2, and 3. He then became flight commander of basic training until Course 9, when basic was eliminated and students went directly to advanced, with no decrease in training hours. He served as squadron commander through Course 20 in November, 1944.

Cousins spoke at length on the decision of British RAF senior officers to eliminate the BT-13 from the basic flight segment of the six-month training program and to have cadets go directly from primary training on PT-17 Stearmans to the AT-6 Harvard:

> That was a move that received a lot of back room criticism from the U.S. Army brass. A U.S. Army Air Corps general in the South East Training Command said that the jump from PT-17 to the AT-6 was going to result in a lot of training accidents. He remarked that he'd like to have the funeral parlor concession when the changeover happened.

The general's crystal ball was "socked in" with fog and his grisly, insensitive prediction was wrong on all counts. The British had never been sold on the BT-13 and they experienced no flying fatalities as a direct result of the changeover.

Cousins spoke of the British flying training methods:

> As a matter of fact, the British were miles ahead of us in training techniques and we (American civilian instructors) were happy to train the RAF cadets strictly under the British system.
>
> The concept of the BFTS program targeted one primary goal: to turn out pilots by concentrated training within the proper allocated time. The Arnold Scheme (U.S. Army Air Corps training system) didn't zero in on the real needs of the British.
>
> To begin with, the Army Air Corps did all of those ridiculous things like hazing underclassmen and making them eat at the mess while sitting at attention. Stuff not connected in any way with flying. It's no wonder that the washout rate in the Arnold Plan was so much greater than that of the BFTS. Some of the boys who washed out at Carlstrom got a second chance and were reassigned to Riddle. Several of them completed our Course with no problem and went on to become great pilots.

Riddle Field's daily log sheets show something of Jim Cousins's resourcefulness. In early October, 1941, an urgent warning went out that Clewiston was in the path of an approaching severe storm of hurricane velocity. Cousins was called on to take any action that could protect the inventory of aircraft at the newly opened field.

We didn't have any time to waste. We had to get those airplanes up country for safety, but we didn't have enough people to fly them out. What we did was fly over to Carlstrom before first light to borrow pilots.

The problem was that none of them had checked out on BT-13s so we gave them (all primary instructors) the quickest conversion to basic trainers on record. All they got were three rapid takeoffs and landings and we pronounced them fully qualified. Then off they went!

Cousins led the flight of BT-13s up to Tallahassee, the state capital, even though one squadron wag warned against it. "Bad location, Jimmy, one of these days God is gonna get even with those politicians and today might be it." When they arrived at Tallahassee, each aircraft was tied down and the pilots went off to hotels. All except Cousins, who wanted to remain close to the airplanes. He spent the night in the operations hut on a desk. In the meanwhile, Squadron Commander Ernie Smith was working feverishly to locate enough pilots to fly the remaining AT-6s and, finally, the advanced trainers flew off to Ocala, toward a safe haven.

But the hurricane, christened Annie and promised to Clewiston, proved a fickle one, indeed. It veered northward, missing Clewiston by many comfortable miles and slammed into Tallahassee, the exact place where Clewiston's basic trainers had been moved for safety. Cousins, hunkered down in the operations hut, had a front-row view of the damage created by the high winds. He saw an old B-18 (DC-3 bomber version) doing its first and only vertical takeoff, suspended about three feet off the ground with its mooring ropes straining to hold it back. The large landing wheels would bump up and down as the screaming winds raced over the plane's wing surfaces, forcing it into the air and then releasing it in a hard thump on the ground. Miracle of miracles, thought Cousins. Straining his eyes through the window of the hut, he could just make out his BT-13s, all six of them, moored safely with no apparent damage. He and his pick-up pilots had flown 350 miles to avoid a hurricane that never happened and ran into one that wasn't supposed to happen but did. At least the airplanes were not damaged but with all the confusion, it took three days to get them back to Riddle.

Cousins remembers another weather story. At the early stages of 5BFTS there was a big push by the RAF to take advantage of every possible moment to get the necessary flying time out of the way and into the log books in order to deliver much-needed combat pilots to England. RAF Wing Commander Kenneth Rampling, a tall, jovial man who was labeled a "regular guy" by the Americans, always had a burning desire to see Course graduations take place on time with as little lost motion as possible. "Let's move it on, lads," was one of his stock expressions. In the nicest possible way, he spurred the staff on to getting airborne from six in the morning until seven or some-times later into the night.

One morning we had all our BT-13s lined up wingtip to wingtip on the flight line. It was a miserable day with poor visibility and a ceiling of maybe 300 feet. We were all huddled together, muttering about what a lousy day it was for flying. All of a sudden we see the wing commander's car coming slowly down the flight line. The car stops right in the middle and Rampling gets out. He is as cheerful as though we were all basking in the sun.

He walked up to our group and threw out his arms as he took in the sights. "What a grand day," he said, "it's just like home! Why isn't everybody flying on this day?"

We looked at each other and then someone said, "Well, sure, why aren't we flying?" So we fired up the BT-13s, took off in the stuff and we were in business.

Recalling some of the escapades of the high-spirited RAF cadets, Cousins chuckled over some of his memories:

Of course there were a lot of things that went on that we never heard about, but this one day I was called up to the control tower because a pilot was on the radio telling about some trouble.

The reception was pretty bad and we could just barely make out what he was saying. Finally this faint voice came through and I asked him for his location. "I'm about four thousand feet over Fort Pierce," he said, "and I'm in some telephone wires."

Well, it turned out that he had ripped down the wires and was trailing about fifty feet of the wire behind him, but everything worked out all right and he landed without too much trouble.

When it was apparent that the air war tide was turning against Germany and there was actually a surplus of RAF pilots, the British began to start cutting back on BFTS locations and then the U.S. Army Air Corps moved to take over some of the facilities. Cousins recalls how the Air Corps made it clear that it was in a cost-cutting mood:

> The Air Corps had some captain in the contract and procurement section call a meeting in New Orleans with operators of the various BFTS fields to discuss future contracts with the government. Well, that captain was mighty unpopular when they all sat down to talk money matters. The British had taken a "to-hell-with-the-cost" attitude because all they cared about was getting pilots in the skies over Germany.
>
> As all of these civilian operators were drinking Sazeracs (bourbon and absinthe cocktails) during the meeting, things got pretty hot and heavy when the army captain started telling them about how the Air Corps would cut back on hourly costs and other reimbursements. People who were there said that a couple of dishes started flying toward the captain and he excused himself in a hurry and went back to Maxwell Field.

T. L. "Tommy" Teate knew that he had to fly in 1927 when he was only ten years old. Charles A. Lindbergh had instantly attained world hero status after packing a lunch bucket with a couple of sandwiches and flying solo from New York to Paris in his *Spirit of St. Louis:* "That flight lit my fuse, as it did for all of aviation, and I knew that I had to learn to fly so that I just might pilot an airplane across 'the pond' some day."

And that he did. Tommy Teate flew across the Atlantic solo, single and multiengine:

> I delivered newspapers, washed airplanes, cut grass and whatever to get enough to buy a bicycle with enough money left so I could bike fifteen miles to take a thirty-minute flight lesson on a J-2 Cub.
>
> I kept at it and finally got my private license at sixteen while still in high school. I was the proud owner of a Lindbergh scrapbook. There was no way I could go to college (remember the Depression) and, besides, I wanted to fly, so I continued working. I passed my commercial and instrument and began working for Embry-Riddle in Miami as a flight instructor. Then I became chief flight instructor and, later, became a flight commander at the 5BFTS at Riddle Field.

I instructed RAF cadets on Stearmans, BT-13s and AT-6s until 1942 when the United States entered the war and I was commissioned a "lootenant" in the Army Air Corps.

Teate flew with the U.S. Army Air Corps, delivering all types of aircraft worldwide. Then he flew "the Hump" in the China-Burma-India Theatre of Operation, returning to the United States as commanding officer of a B-29 transition school. One of his duties was flying the showcase B-29 Superfortress to war bond rallies in cities throughout the country. Everyone wanted to see the largest, most sophisticated aircraft ever built, and so millions of dollars were raised during the appearances of the "airplane that would win the war."

Teate told of how he received orders to fly to Detroit for a showing of the B-29, with Willow Run Airport as his base of operations:

I went to get something to eat at the cafeteria at the huge Ford factory complex where B-24 bombers were being built. The cafeteria was practically deserted because of the hour. I went through the line and took my tray to an empty table (most were) and sat back to have my lunch. I noticed a few people coming in for a late lunch, as I had done. One tall, rather slim man entered, took off his coat and hat, hung them on a rack, went through the line, and came over to my table. "Do you mind if I join you, major?" he asked.

"Please do," I said.

He began eating and then said, "Major, are you the commander of the B-29?"

"That's me," I said, and then almost fell out of my chair when I recognized my lunch companion. "Are you Colonel Lindbergh?" I stammered. "Yes," he said, "I'm afraid I am." (I learned later that he was a consulting engineer at Ford.)

It took a few minutes for me to compose myself and tell him what a pleasure it was for me to meet the man who was responsible for my being a pilot, flying in a Superfort and having lunch with him.

Lindbergh laughed and said, "I've heard so much about the B-29 that I would sure like to look it over. Do you think you could arrange it? I'll bet it would take an act of Congress."

My mind started racing, and then I said, "How would you like to fly it?" Then I told him to meet me in base operations in a flight suit. But could I deliver the goods? I went directly to a telephone and telephoned none other than General "Hap" Arnold. Would he remember that I had been a pilot for him and some of his staff? I got through

to the general quickly and he gave me the go ahead to take Charles
Lindbergh anywhere he wanted to go on the B-29.

Teate realized his lifelong dream that day at Willow Run Airport.
He insisted that his boyhood hero take the left seat on the B-29, the
pilot's position, as they took the giant airplane over the Detroit area.

The takeoff, the climb and airwork were real smooth. Lindy (it was
OK to call him that by that time) said that it was the largest aircraft he
had ever attempted to fly and, if I recall correctly, the first really
pressurized one.
 After the altitude run of about 40,000 feet, we came down and shot
takeoffs and landings and I honestly felt that he probably could have
made a circuit without me.

Teate flew in World War II, the Korean War, and the Vietnam War.
As of 1994 he was still flying and instructing multiengine and instrument, primarily. He is chief pilot for a group of volunteer pilots who
fly missionaries, doctors, and church groups on humanitarian efforts. Known as Missionaire, Inc., the group gives help and comfort
to needy residents on the islands of the Caribbean.
 Teate retired from the United States Air Force as a lieutenant
colonel. He was a commercial airline captain, a test pilot, veteran of
three wars, and flight instructor for the British Royal Air Force, the
United States Air Force, and the Brazilian Air Force. *He did all of this
as a high school graduate.*

Old friendships don't fade away very soon, and sometimes not at
all. Many former British cadets and their American instructors have
continued communicating with one another and in many cases visiting in person for more than fifty years.
 Former Course 3 cadet Bob Richardson, of Newton Mearns, Scotland, once again flew with his Riddle Field instructor, Bob Hosford, in
1993, during a reunion of 5BFTS Association members in Clewiston.
Richardson demonstrated that "auld" acquaintances need not be forgot when he returned to his native heather and wrote a tribute to
"auld lang syne." Richardson's essay is not only a remembrance of
things past, but is also a tribute to the ability of man to still enjoy those
skills acquired so long ago:

The time is around 10:00 A.M.; the surface of Lake Okeechobee, four thousand feet below, is reflecting the colour of the azure sky. But to the south little wisps of cumulus clouds are forming with thermals rising from the Everglades.

The yellow painted wings of my Stearman PT-17 shine in the strong morning sun and this sturdy biplane is cruising at a steady 95 mph, its 220 hp engine giving a healthy exhaust note at 1,800 revs.

There's the straight road to Fort Myers up ahead and so I start the right hand turn to bring the heading from 180 degrees to 270. Lined up with the road below, throttle closed, engine just ticking over, ease back on the "stick," wait for the shudder of the "stall," kick left rudder and the ground below starts to revolve. The road comes round, half a turn, one whole turn, next half turn kick on right rudder, "stick" forward and the spin stops dead in line with the road and we are diving away! Liked that one—a two-turn "precision" spin!

But this is not 1941. It is 1993 and today is September 24, almost fifty-two years to the day when I first set foot on the airfield which was to be my home for six months. And I am not flying a Stearman. I am piloting a twin-engine Beechcraft Baron with my friend and former instructor, Bob Hosford, who put the control column into my hands on the leg of the flight south from Orlando to Riddle Field, just on the edge of the Everglades.

I swing the Baron into a right hand turn at 1,000 feet over the Fort Myers road, with a sidelong glance at the altimeter. Yes, the needle is spot on the 1,000 mark. Fly down that road a bit, then north to Moore Haven over the junction where that old "juke joint" used to be. I do a "180" over Moore Haven, a left-hander with at least 45 degrees of bank. Wonder how Brenda (Mrs. Richardson) is feeling sitting right behind me in the cabin. First time she's been airborne with *me* flying the aircraft. Some back pressure on the "stick" to tighten up the turn—getting a bit of "G"—this is more like it!

Brenda and I have flown down to Clewiston from Bob and Martha Hosford's Pennsylvania farm for a reunion of former RAF student pilots of Riddle Field.

Unfortunately, we had to fly back to Pennsylvania a little earlier than planned, as hurricane-type weather was beginning to form in the Gulf (just like the old days). But the girls enjoyed the trip. Martha gave us a vote of confidence. "We were doubly safe," she said, "with two Transport Command pilots up front."

Eric Carlson was an instructor at Carlstrom Field early in 1942, just before Carlstrom's training of RAF cadets ended and all new British cadets were assigned to Riddle Field. He then instructed at Riddle, starting in November, 1943. Carlson's first class (Course 18) had both RAF and U.S. cadets. From then on until the final class (Course 26), only British cadets were trained at Riddle.

Early in 1941 I was instructing in a Civilian Pilot Training Program in the New York City area. I had a student in a primary flight course who told me that he had washed out as a U.S. Army Air Corps cadet and could never again receive flight training from the army. The student and I had what they call today an "attitude problem." We simply didn't get along very well and angry words were spoken. The student did complete the course, and perhaps I could be faulted for most of our differences. Anyway, we parted company and I never thought I would see him again.

I then became an instructor at Carlstrom Field, in Arcadia. During midsummer, 1942, a new class of Army cadets came in and about 100 of them were assigned to our flight. The cadets marched over to our ready room to be assigned to instructors, and here was my pupil from New York, the one who had washed out of cadet training with the Army. We made eye contact right away and I was a bit relieved when he wasn't assigned to me.

A few days later on the flight line, we passed each other and stopped to say a few words. "Are you going to turn me in?" he asked in sort of a stammering manner. "I'm not going to squeal on you," I said, "as far as anyone knows here I just met you a couple of days ago when you came on base." Boy, was he relieved. I gave him my address in Arcadia and told him to stop by so we could talk, as I was curious to learn how he got back into air cadet training after begin bounced out.

"Right after Pearl Harbor, I figured I could take a chance and get back in," he told me at my home. "Of course, I couldn't tell 'em that I had washed out of pilot training. I enlisted for cadet training as a navigator, but someone at preflight said I was pilot material, so here I am. First they kicked me out and now they want me back, but they don't know I'm the same guy."

I never told a soul about the cadet's secret, not even my close friends. At any event, he completed the course in Arcadia in fine style and I received several letters from him, telling me that he got his commission and inviting me to his home after the war. But as things happen, we lost track of each other and I don't even know if he's alive today. He changed from being a difficult student to a natural pilot. Sometimes you can't figure it out. I had a few other cadets who seemed to learn very, very fast.

Carlson has dozens of letters from former cadets that he trained either at Carlstrom or at Riddle Field. Here are a few excerpts:

April 3, 1992
I have no doubt that this letter will be a surprise to you, coming about forty-seven years late. My wife and I returned from Florida a few days ago after a most enjoyable three weeks' holiday.

It was a trip down Memory Lane when we visited Clewiston. I was in Class 22 and I first flew with you as my instructor on 18 December 1944.

On 24 January 1945 I made that terrible mistake and landed with my wheels retracted on an auxiliary field. Anyway, I survived the check flight with Flight Lt. P. J. Stephens and was allowed to continue on the course. However, I know that you put in a very good report, which helped a great deal. If I didn't thank you then, I do now.

I had another dreadful mishap on 19 March 1945 when I nosed up an AT-6 trying a short takeoff. The aircraft unfortunately caught fire and burned. Helped by another good report from you and a successful check ride with Flight Lt. Stephen Harvey, I again survived and got my wings on 31 March 1945. Now, perhaps you remember me!
                    Alf Bell
                    Durham, England

11 January 1992
Turning the pages back forty-six years—we got our wings two weeks after V-J Day and traveled home on the Queen Elizabeth *to find that the RAF considered us all surplus. As we were now "square pegs," they did their best to fit us into round holes.*

After three months in the UK, I found myself on my way to India, spending a week in Cairo and then reaching my station in India, which was to be Agra, home of the Taj Mahal. My role was to be catering officer to many RAF and Indian Air Force personnel. This was to be my lot for eighteen months while we handed India back to the Indians. On leaving the RAF I started a new career with Harrods department store and spent thirty-eight years with them in an administrative capacity.
                    Alan Boyd
                    Devizes, Wilts., England

May 17, 1943
I suppose you can still remember the raunchy cadet who had you sweating every time a check ride came up. I went to B-24s and after

riding a long time as copilot went overseas, and am now a first pilot.
I can't tell you much of what we are doing or where we are. I am in
the middle of the Eastern Theatre and have quite a few combat missions
under my belt. We give a hell of a lot of credit to you for all of us
getting through. We learned a hell of a lot of flying down there and
would like to hear all about Carlstrom and your present batch of
students.

Lt. Bill Underwood
APO 683 New York City

June 22, 1943
Well, I am about through basic and I don't know what I'll take next,
twin engine or single engine. My choice has always been bombers, but
about two weeks ago I put single engine as my first choice. I may be
able to get it changed at the last minute. As far as I know, I'm the only
student down here who hasn't "buzzed" anything yet. I think I was
taught very good habits in primary at Carlstrom.

I would like to be up north, as I have never seen such hot weather as
down here. Another reason for wanting to get away is the women
situation. I have heard a lot about Georgia "peaches," but they must be
the ones that grow on trees.

Ralph Howell
Bainbridge, Georgia

April 6, 1944
Well, here I am writing to you at an altitude of 4,000 feet over the
jungles of South America. Departed the States last Friday and have
covered the territory ever since. I am taking a C-47 places or, rather,
George (the automatic pilot) is taking us places. It's a lot different, this
type of flying. Just moving along and keeping up with my
correspondence.

Lt. Bill Hendrix
Memphis, Tennessee

February 12, 1944
This is a long overdue letter. In case you've forgotten me, I'm the cadet
from 44C who had to be talked into flying. I'm glad you talked me into
it because now I eat the stuff up. I feel I can really fly now because
of you.

A/C John Silverman

13 August 1944
I am pleased to say that we all returned home safely and were packed on
home for a fourteen-day leave. My girlfriend Mary and I are now
engaged. Right now I am stationed at a small air base in Yorkshire . . .
before going on to twin engine machines. I would like to express my
sincere thanks to you for your great part in my training.
                    Sergeant F. G. Cocker
                    Yorkshire, England

13 June 1944
Since arriving back in England, quite a number of fellows from Course
21 have been sent on a flight engineer course and I was amongst these.
We have finished the course and are now awaiting posting for some
flying on Lancasters. The weather here has been pouring rain for days
and has made me think longingly of the Florida sunshine. Thanks for
persevering with me—a pretty dim pupil—while I went through my
"teething" period in flying. I certainly shan't forget your help.
                    Sergeant Philip Dare
                    Devon, England

In September, 1945, with training duties ended, Carlson returned
home to New York to his wife and 18-month-old daughter. He did
some flying instruction, in addition to other jobs, and went to work
for a laboratory division of a major chemical firm until retirement in
1979.

Alex (Dusty) MacTavish was an instructor at both Carlstrom and
Riddle Fields and was a personal friend of John Paul Riddle. Still
flying in his seventies, MacTavish has a Czech-made Zlin, a vintage
low-wing stunt plane that he flies from his Lake Worth, Florida,
home. Fit and trim enough for a man twenty years his junior, he is
as at home on the golf course as he is in the cockpit, and managed a
golf course for several years in the New York City area. Born in
Canada of Scottish immigrants, MacTavish followed his father to
Florida as a youth and spent most of his life in the southern states.
  He spoke about his flying experiences:

As far as I'm concerned, flying is a lot like playing the violin or any
other musical instrument. You've got to do a lot of things and think a
lot of things all at the same time. The trick is that you can't sit around

and decide what to do next; that violin or that airplane has to be an extension of yourself.

I got some real breaks when I was an instructor for John Paul Riddle because the cadets I drew didn't get into any real difficulties. I can't remember any of them even ground looping a Stearman, something that was quite a common thing. Maybe it sounds like I'm taking more credit for my ability as an instructor than I should, but my groups got along pretty darned good.

Oh, there were those times that some kid would do some dumb thing, but nothing major. If you pointed it out to him in an understanding but firm way, you'd get his attention ninety-nine times out of a hundred.

I think that one of the keys to being a good pilot is to have better than average athletic skills. Don't get me wrong. You don't have to have the eyes of a Ted Williams or Babe Ruth, but it comes in handy if you took part in sports in school. British boys who were pretty good soccer players frequently had a real edge in primary training.

When I started flying as a kid, I had only an hour and forty minutes of supervised flight time when I soloed. But that was nothing. I found out later that the guy who gave me instruction didn't have any license at all. But I was flying every day, and along the line got to thinking that I was pretty darn good.

I was building up hours daily when the war came along. I figured that I could do some good in training pilots. That's when I went to Carlstrom Field and signed on as an instructor.

One thing you've got to say about primary training back in those days is that the Stearman was one hell of an airplane. It could take just about anything you could give it. All in all, it was one of the best training planes ever made. Yeah, we did a lot of aerobatics with the students and all sorts of things. You always had to figure that the engine was going to cut out when you rolled it on its side, but, hell, it ran again when you rolled it back over.

We could climb up to a high enough altitude and try to do lazy eights. The engine—exactly as expected—would cut out, but nobody would get too excited. When we were gliding down, we'd get it restarted. There was really no sweat. The Stearman was tough enough, but didn't have the power to do an outside loop.

My students used to get a kick out of hearing about crop dusting. I used to tell them how it was a piece of cake before electricity found its way down south and pretty soon every farm had wires strung out along every damned cotton field.

A crop duster would think he knew every square foot of land

down there, and all of a sudden, he'd be flying over a farm and bring down the plane to dusting level when wires that weren't there yesterday popped up in front of him. I lost count on how many forced landings I had to make because of those wires. I tell you that when "civilization" started taking over the South Carolina farms, it made it tough on us.

This one time I was in a Piper Cub when the engine started sputtering and it looked like it was going to quit cold. I looked around for the usual emergency landing spot—a road—and I was watching out for power line poles because you couldn't see the wires until it was too late. If you could spot the poles, you had a good chance of ducking under the wires.

That was the theory, but this day it got all shot to hell. I was watching the pole and a hedgerow on the other side of the cotton field when I hit another wire that crossed the field. One end of my prop was knocked off and the plane started shaking like crazy. It didn't take long for the unbalanced prop to cause the compass to be shaken right out of the instrument panel. A fuel line ruptured and gasoline started spraying all over me.

I managed to pull up and shut off the engine and then I glided over to a straight stretch of dirt road and landed. The plane no sooner stopped rolling than I was out of there, running like hell. When I went back to check out the plane I found that the engine mounts were broken. In another few seconds, that engine was bound to have dropped off.

Dusty MacTavish was one of several crop-dusting pilots who trained British and U.S. cadets at Riddle and Carlstrom Fields. Many cadets who won their wings and went on to log many hours both during and after the war have high praise for the skills passed on to them by those seat-of-the-pants fliers who plied their chancy trade throughout the South and Midwest. In a personal interview, MacTavish tells of a frightening, yet humorous incident that usually gets top attention whenever a group of seasoned pilots get together for a bit of "hangar flying":

I was doing some dusting up in South Carolina and dropped down from the north end of a big cotton field, only to find a big new power line right in front of me. All I could do was give her full power and hope to cut right through it, but when I smacked into it, it just kept stretching like a giant rubber band for what seemed like two miles.

My plane kept straining as though I was on the business end of a slingshot. When the wire stretched as far as it would go, my plane was shot backwards and I ended up in the top of a peach tree. I wasn't hurt at all until I climbed out of the plane, stepped on a broken limb, and fell out of the tree.

It was then that I figured there were plenty of other flying jobs that a young fellow might consider.

This account is from a former Clewiston instructor, living in New Orleans, who told us his experiences with the understanding that it be used without attribution:

When Mr. John Paul Riddle notified me that I had the job, I felt like jumping up in the air like kids do and screaming something like "yippee." I was that excited and relieved after being turned down by just about every flying school in the country.

The envelope was postmarked Miami, Florida, October 13, 1941, and the letterhead was printed in blue ink: Riddle-McKay Aero College. The letter was, as we used to say in those days, "short and sweet."

I carried that letter for probably twenty years or so in my wallet. A wallet would wear out and "The Letter" (I had given it that title) would be moved to a new wallet along with driver's license, Social Security card, Blue Cross card, and all the other stuff that you need to keep on living in this country.

"The Letter" finally was consigned to a safe deposit box, where I keep all kinds of things that were very important to me at one time or another: a valentine from Nancy Miller when we were in the ninth grade; a piece of Chief notebook paper carrying the autograph of Jimmy Doolittle; a few letters from former Brit cadets who had me for advanced at Riddle and a copy of the Royal Air Force pilot training manual.

The gist of the letter from Mr. Riddle was that I could start work as soon as I could get down to Clewiston and that he felt sure that I would "have no further problems" that would concern either him or his staff.

What he was driving at was a reputation (undeserved, I felt) that was sticking to me wherever I went as a flying instructor. That I was doing too much "bar flying" was the scuttlebutt that went from one field to another. Like most rumors, it had some truth to it. When my short-lived marriage blew up, I may have gone on a few binges, but never drinking just before flying and never flying with a "morning after" hangover that would make me sloppy at the controls.

I leveled with Mr. Riddle when I wrote for a job and I didn't lie or try to sugarcoat what had happened to my marriage or try to put the blame on my ex-wife. As far as drinking was concerned, I didn't say that I was on the water wagon (I wasn't) and I didn't try to label all of the rumors about me as bullshit. At least *most* of them were.

All I asked for was a chance to join his group and help train the British pilots who were coming into Florida. Then I wrote something that always bothered me because it was taking advantage of a friendship.

My father once served in a civic organization with Mr. Riddle, and they had developed friendship and mutual respect for each other because of that association. My letter asking for a job closed by saying something about "you know my family enough to realize that I would never do anything that would discredit your organization." It wasn't my finest hour when I wrote those words, but I was just as short on pride as I was at the bank. Out of work for four months, I was renting a third-floor room in Alexandria, Virginia, and I had forgotten what a paycheck looked like.

The letter from Mr. Riddle jump-started me out of feeling sorry for myself and made me think that maybe I did have something to offer and that I could regain something that I had lost along the way during the last couple of years. I was a good pilot and I was young, only twenty-five.

Two days later, I was on an Atlantic Coast Line train heading for Clewiston, Florida, where I was to report to a fellow named John Cockrill to get the lay of the land.

I knew I had more flying hours than many of the instructors, although some of the real old-timers had put in an amazing amount of flying time. "I've got more flying time than you've got sack time," Dusty MacTavish would frequently say to me in a good-natured way. I was almost seventeen when I started flying at Cleveland Airport because my father knew I wasn't cut out for a long ride in college corridors. He figured that airplanes were here to stay and maybe there would be a place for me. When I hit twenty-four, I had seven solid years of flying behind me, more than 2,000 hours.

I got into Clewiston sometime on a Thursday afternoon and went to the Clewiston Inn. I got a room and bath for about two dollars a day. I don't think it was any more than that. Later on I found that most of the single instructors were renting rooms in private homes for anywhere from twenty-five to fifty dollars a month.

I met John Cockrill at Riddle Field on the day after my arrival. He was expecting me and went over my credentials, commenting quite favorably on my experience as an instructor and taking what was obviously a genuine interest in wanting to help me adjust to life at Clewiston. Cockrill was both tough and fair-minded, a guy who treated his pupils with understanding and his subordinates with respect. I know it's an overused expression, but no one could better fit the term, "a flier's flier."

That was John Cockrill, cowboy boots and all. I felt down deep that he knew about the problems that I was supposed to have had. I am just as certain that Cockrill kept the information to himself because he was not the sort of man to mouth off stuff about anyone.

I don't remember how many Wings Parades I saw during my time at Clewiston. I left some time in August, 1942 to go on a special training assignment for the Army Air Corps cadet program that was supposed to be "top secret." John Paul Riddle called me the day after I got the offer and told me I had done a great job and that he was proud of me. I later found out that Mr. Riddle had given me a top recommendation when he was contacted. My personnel folder was open on a desk and I took a look, figuring that no one told me not to peek. A notation stated: "sorry to lose his teaching abilities, but feel that he can make many more contributions to the war effort in this assignment." It was signed by John Paul Riddle.

During my relatively short time at Clewiston—just under a year—I remember that two British cadets were killed in night flying crashes. They were Roger Crosskey, flying a BT-13, and Richard Thorp, some months later in an AT-6A. With all the young guys sweating out ground school and pushing themselves so much every day, it's a wonder that more mistakes weren't made. I remember a strange incident that bears this out. I was flying solo one day when I spotted a Stearman on the ground in a field about ten miles or so from the main airfield. It turned out that an instructor and his student had engine trouble and made a forced landing in a swamp. The field had six to eight inches of water and on top of this was a lot of floating grass and reeds that made it appear like solid ground from the air. While I was figuring out directions and distances to tell a recovery crew at the base, another Stearman appeared, entered a "traffic pattern" and landed alongside the first one in a cloud of spray. The water was too deep for either of them to take off, so they had to be dismantled and trucked out of there. The second plane was piloted solo by a cadet who said he saw the other plane down there on the ground and thought it was an auxiliary field.

I found that all the instructors did everything they could to make me feel at home. If they knew anything about how I got the job, they sure didn't let it show. Charley Bing and Paul Greenwood were good company, as were several others.

When I left Clewiston for the Army Air Corps assignment at Maxwell Field, Alabama, I really had a feeling of appreciation for all that had been done for me after going through some pretty rugged times. I stayed on the job with the Army Air Corps until the end of the war without ever learning what was so damned secret about what I was doing.

As the Brits always said, "it was just the Air Force Way."

That must have been it.

Giving your instructor "the business" and watching him squirm—as he had made you squirm countless times in primary training—was the dream of many a cadet. All during that initial training stage, starting from day one when you take your first dual instruction with a guy who has been flying since birth (and don't you forget it, pilgrim), you have but one thought: "I could kill him, and no one would care. Not even his wife or kids. Even his mother would just shrug."

You had to listen to enough "do's and don't's" in a few weeks to last you the rest of your life and if he didn't get off your back, the rest of your life wouldn't be far off. One day he got out of the Stearman and told you to go it alone. "You seem ready as you'll ever be, pal." And why not? You had been pushed and prodded to the breaking point and now were ready to solo. And solo is what you did, with the Brit version of a "whoop and a holler," as you took that baby around, all alone up there, and wasn't it a kick and a half? When you landed—was there ever such a landing?—"he" was there with his hand stuck out and a big grin on his good-looking face. There, that wasn't so tough, was it? He actually smiled and shook your hand. Hey, he's not so bad after all. From then on, all during primary, it was a love-hate relationship. The instructor didn't exactly get off your back. You got a dressing down when you flubbed, but at least you believed him when he said the same old thing over and over: "Hey, look, I'm not your mum and I'm not going to coddle you, but by God I'm going to break my neck to get you into that Wings Parade and I'm going to do my damnedest to see that you don't break *your* neck. You're a pilot now so you gotta act like one."

Primary is almost over and now comes the big traditional day. Cadet Desmond Leslie, Course 5 at Riddle Field, could hardly contain himself, as he describes it in a March 1, 1942, report:

Well, here we are at the end of our primary training course. All we have to do now is three hours of formation flying and then we go on leave. Now we are allowed to take our instructors up for a "lesson." The controls are on the other foot and hand. What a classic event!

I start by putting my instructor under the hood and having him take off on instruments, a thing we all had to do. As he was the pupil, he followed my command by saying, "Yes, sir!" I was to call him, "Hey, Walker."

He did a good instrument takeoff and I told him so:

ME: "OK, Walker, that was OK. Now do climbing turns and level off at 3,000 feet."

Another plane approached so I took over for a minute or two, as he was flying blind.

ME: "When I say OK, I want you to take the controls and recover straight and level flight at the end of the maneuver. Do you understand?"

WALKER: "OK, do your damnedest."

ME: "Watch your instruments."

Then I did a loop, a slow roll, a quick turn to get his compass nicely messed up. Then a one-and-one-half snap roll and before he managed to figure out what was happening, hauled back on the stick and we went into a split-S and into a steep dive.

ME: "OK, Walker, that was a good recovery. When I say recover, do the same again."

I became aware of a hand sticking into my cockpit, holding a little note on which was written "no inverted snap rolls." Apparently he thought I was trying to shake him out of the plane.

ME: "This next one is the Juke-Joint Right, invented by the pupils of this school to frighten the Japanese. OK, Walker, come out from under the hood and be sick."

WALKER: "Please, sir, may I make you sick now?"

ME: "Go ahead and try."

Then I sat back, tightened my belt and waited. He did three outside snap rolls, a flick at 110 mph, a stall off an Immelmann, and a power inverted spin down to 1,000 feet. In my best oh-so-suave David Niven manner, I tried to look painfully bored in the mirror.

ME: "Well, Walker" (I drawled). "Are you flying this plane or is it flying you?"

I made my mistake in throwing back those cutting words that he had often used on me. Wham! I was cut off by a vicious jab in the stomach as he pushed the stick forward in a steep, fast dive. He pulled out just before the blackout point, almost pushing me through the floor.

WALKER: "How did you like that, sir?"

ME: (fighting for breath, but sticking to my "instructor" role): "Walker, your coordination is rough, your maneuvers are indecisive. In fact you are just plain bad."

WALKER: (such a penitent "pupil"): "Oh, I am so sorry sir. I just don't seem to know how to make it. Will you show me how, sir?"

ME: (hastily pulling back the throttle): "To practice a forced landing, select that field with the palm trees in the corner."

We glided down and I turned off the ignition. At about 20 feet, I said, "OK, open her out and fly home." The next few seconds were a treat as he opened the throttle and nothing happened. I switched on as we were about to land and he roared with laughter and relief. "Time's up, Walker," I said gruffly, "take me back to the field." After we were finished with our game, I solemnly wrote out a "grade report." "I am afraid I can't pass you on, Walker," I said, "I'll have to get Mr. J. (another of his pupils) to check you 'round again."

We then saluted each other and dissolved in laughter. What a great ending to our course! Meanwhile, in the distance, we could see other chaps giving their instructors forced landings in swamps, making them do circuits of the field, and all the other "grinds" that they had inflicted on us during the past three months.

We could see what a grand day it was and how the students and instructors enjoyed it all. To cap it all, the commander said, "Let's all fly back to the main field in formation." All together? We had been flying in small formations, but now off we went—all thirty aircraft—in one glorious straggly V. Somehow, we managed to get in the air together and we did a grand sweep over Clewiston to the chagrin of the senior flight. They had put up a nine-ship formation that morning for our edification. Now they had to witness the junior class with more of our Stearmans in the air than they could muster with their noisy silver Harvards, which made such a racket.

I shall be sorry to leave these reliable little Stearmans; we have had such fun with them. Next week shall be momentous. I shall send bulletins from New York. A week's dissipation will do us all good after months of hard work and nervous strain. Flying is a hell of a sweat until you get used to it and are not afraid of the plane. I was really scared stiff at first, but with a good instructor fear quickly passes.

～ Community Friendships ～

There was no doubt that the Arcadia and Clewiston girls found the young British cadets light years ahead of the American trainees when it came to "dateable targets," and it caused some resentment among the U.S. fliers. Most of it resulted in fairly good-natured kidding or digs directed toward the Brits.

"Hey, Reginald old boy," a Georgia trainee would call out, "you and that little gal seem like mighty serious stuff. You gettin' any of that?" But for the most part, the socializing between cadets (Brits or Yanks) and the southern girls mainly amounted to little more than dancing at a church hall or having Sunday dinner with the girls' parents. A Royal Air Force cadet would be expected to amaze his date's younger brothers with tales of reckless courage and valor. ("There I was at five thousand feet, upside down, when the engine conked out . . .") The girl's mother, accustomed to *Modern Screen* gossip about Hollywood royalty, just knew that the young man with the fascinating "brogue," "the spitting image of Leslie Howard," must know as much about the goings on at Buckingham Palace as *she* did about the movie star shenanigans at MGM or Paramount. "I can't for the life of me understand what that Edward ever did see in that Wallis Simpson. She was no better than she had to be, I say. And putting on like she was southern and all the while she's from Baltimore. Might as well be from Boston." All the while the American matron is holding forth, "Leslie Howard" is cutting into his roast chicken and sage dressing, expertly wielding a knife with his right hand and heaping chicken and dressing on the back of the fork's tines. "Oh, Momma, don't you love the way he does that? Don't you just love it?" the girl gushes. "Just like Greer Garson," the mother says. "She used her silver just like that in *Goodbye, Mister Chips,* exactly like that."

Maybe, just maybe, the cadet and his date would have a chance to be alone on the porch swing or take a stroll in the back yard if the mosquitoes weren't too ravenous, and maybe they would kind of edge into each other in sort of a pretend stumble ("And, oh lord, he reached up and touched me and I swear that he squeezed up here and I thought I might have to push his hand away unless he stopped in a little while"). All pretty tame stuff even by the standards of the day, standards in 1942 that ranged from light petting ("don't mess up my hair and I don't French kiss") to Errol Flynn bodice-ripping stuff and "going all the way" with a two-dollar, ten-minute date above the Puritan Ice Cream store.

British cadets indeed were objects of curiosity, to put it bluntly. The America of the 1940s—or at least the back country of America— was pleasantly surprised by the sizable influx of smartly uniformed visitors who were going to be in town for "quite a spell." (A typical comment by local Floridians: "They aren't really foreigners, Charley. They speak English as good as those Yankees who came down to work over at the air base.")

The Florida towns of Arcadia and Clewiston, each boasting populations of about 3,000, could well have been *Life* magazine models for prewar Smalltown, U.S.A. They were busy towns and like other communities in the rest of the country, were climbing out of the sinkhole of the Great Depression. Sugar was king in Clewiston, home of the U.S. Sugar Corporation, while large cattle ranches dominated Arcadia.

The training fields near the two towns were welcomed by almost all rank-and-file citizens, but did provoke some criticism. There were some whispered negative opinions among business interests who predicted that employment opportunities at Riddle and Carlstrom Fields would drain off workers from the traditional sugar and cattle labor pools. The comments were for the most part confined to such occasions as a business lunch at the Clewiston Inn or a service club meeting in Arcadia among persons of similar interests. Before bringing up such a labor shortage possibility, one was careful to lead off with something like, "God knows I'm all for the air field, but . . ."

Most of those fears proved groundless, even though U.S. Sugar did face a severe shortage of help during the 1942 sugar harvest. Production of the cane harvest was crippled primarily because agricultural workers were lured away to booming war production factories in other sections of the country. Civilian employment at the flight training fields chiefly consisted of persons drawn to the fields because of special skills.

At the classes of British cadets continued to flow into Riddle and Carlstrom, the warm links between the airmen and townspeople grew stronger and, quite frequently, touching. Many cadets felt the pain of separation when they were sent to new posts after graduating or (sadder still) when they washed out of flying training. More than half a century after the British flying training programs came to a halt in the United States, many persons still marvel at how quickly the British cadets fit into the social life of south Florida. One reason was that they had an opportunity to learn about a new land and what an almost completely different way of life had to offer. And then there was the way the townspeople reached out ("These kids are like mine!"), and that feeling was returned by the young homesick Britons ("They're like Mum and Dad!").

When British cadets at Carlstrom had more than overnight free time, they would frequently go from Arcadia to Sarasota, Bradenton, Fort Myers, or Tampa, all along the Gulf of Mexico. Riddle Field cadets at Clewiston did visit Sarasota quite often (the circus was a big draw), but more than likely when away from Clewiston they would be inclined to go to Palmdale or Moore Haven or east to Belle Glade, Palm Beach, West Palm Beach, or Miami. Wherever they went they found gates, doors, and hearts open wide to receive them. Florida and the British cadets were made for each other, it seemed.

A Very Special Person Makes Palm Beach Home to Boys

She had grace, dignity, and compassion, not to mention a compelling mission to make life a little easier for (in her words) "those British boys so far from home." She also had the social connections in Palm Beach, playground of the "old money" millionaires, whom she encouraged to join in her pet project. This was Mrs. Ira Nesmith, a Canadian national and wife of a prominent banker, who had lived on Palm Beach's exclusive Chilean Avenue since 1936. Florence Nesmith used her influence ("ladylike clout") to provide recreational housing for British cadets who flocked to the Palm Beach area during free times. With their own money and contributions from fellow members of the Everglades Club, the Breakers

Beach Club and other individuals, the Nesmiths provided hospitality for more than two thousand boys while the British flying programs were in place during 1941–45.

Florence Nesmith thought nothing of calling social register lumi-
naries for contributions to what friends jocularly labeled the FFFF,
Florence's Friendly Fund for Fliers. But contribute they did, and
very willingly. Most of the Palm Beach and West Palm Beach residents who joined Mrs. Nesmith were private persons who avoided
the limelight. Some were members of the English Speaking Union
or the Bath and Tennis Club or parishioners of Mrs. Nesmith's
church, Bethesda-by-the-Sea Episcopal. There were Clevelanders
Mrs. Dudley S. Blossom and Mrs. Chester Castle Bolton, both of
South Ocean Boulevard, and Mrs. William H. Buckley of New
York, and "Kathwill," 131 Sea View Avenue.

Then there were those who regularly were seen in the Palm Beach
haunts, hounded by photographers, and with names that fre-
quently popped up in Walter Winchell's syndicated gossip column.
John Jacob Astor V reportedly signed a check and told Florence
Nesmith to fill in the amount. He was the scion of one of America's
most fabled businessmen, John Jacob Astor IV, who went down
with the *Titanic* in 1912 after putting his pregnant wife on a lifeboat.
Astor, who died at seventy-nine in 1992, had graduated from St.
George's School, Newport, Rhode Island, in the 1930s. Sir Harry
Oakes and Lady Oakes lived at "Seven Oaks," 131 Barton Avenue.
Oakes, the Canadian gold millionaire, was a regular contributor to
the fund in the very early days. He became the center of bizarre
tabloid stories on July 8, 1943, when he was shot to death in his
sleep at Westbourne, his mansion outside Nassau. His body was
discovered after he had failed to appear for a golf date with the
Duke of Windsor, governor of the Bahamas.

Florence Nesmith had an equal opportunity policy on donations.
Her favorite nephew, a flier in the Royal Canadian Air Force, had
been killed in action, and she dedicated her time, money, and effort
to provide as much comfort as possible to those cadets who soon
would face the horror of war.

One of the larger parties planned by Mrs. Nesmith for Royal Air
Force cadets was a British War Relief Society benefit. More than 500
Palm Beach residents assembled at the Bath and Tennis Club, a

landmark of exclusivity. They were there with checks in hand to swell the fund, which, by late 1941, had financed the shipping of some 30,000 large packing cases of warm clothing to Britain. (Whenever "serious money" donations were concerned, the "in places" to fill the coffers were the Bath and Tennis Club at the southern end of Worth Avenue or the Everglades Club at the northern end.) The Princess Dona Marta de Habsburgo-Borbon, an enthusiastic booster of Florence Nesmith's successful efforts, was ticket chairman for the posh gala. The princess, known internationally as the Archduchess Franz Joseph of Austria, was a tireless worker for the British cause and had no difficulty in drawing a large turnout for the fundraiser. She and the Archduke Franz Joseph (Prince Don Francisco José de Habsburgo-Borbon) maintained residences in Palm Beach, New York City, Cannon Mountain, New Hampshire, Austria, Italy, and Spain.

Very much the center of attention were slim, trim RAF cadets who had weekend passes from Riddle Field. Dashing in their "blues," the flight trainees charmed the affluent Bath and Tennis Club crowd during the cocktail hour, telling how they looked forward to "getting a crack at the Hun." All the while they managed quite nicely to reach out (usually without sloshing their martinis or manhattans) toward heaping plates of shrimp, assorted cheeses, and thin bits of toast topped with caviar, anchovies, or ham salad. (It is interesting to note that the Bath and Tennis Club, a bastion of white Anglo-Saxon Protestant members, banned all Jews and blacks from membership and even from entering the club as guests. However, this rule was relaxed during the war years. Royal Air Force and U.S. servicemen with "Jewish" names were accepted without question as guests at both the Bath and Tennis Club and the Everglades Club, the island's oldest.)

The early evening at the club was devoted to the showing of three films, two short features of Scotland and London, and the Alexander Korda morale-builder *The Lion Has Wings,* a 1939 seventy-six-minute "docudrama" dramatizing Britain's entry into the war and the strength (exaggerated) of the RAF. In 1942, the Hungarian-born Korda received the first British knighthood ever conferred on anyone in the film industry. The Palm Beach preview of the unabashed RAF film tribute brought wild applause and loud cheers as it faded to the end. Five hundred anglophile voices, plus those of two dozen RAF cadets, joined in raising the roof with an energetic *a cappella* rendition of "God Save the King."

Following the film, guests mixed for a time and then had dinner on the patio at tables placed beside the pool. Buffet tables were in the dining room. The RAF cadets were seated individually at scattered tables so that they could be available for conversation with a cross section of guests. An entertainment program followed the supper, with the popular baritone Jackson Hines accompanied by the piano duo of Pliner and Earl, headliners at the Hotel Royal Worth. Jackson became more of a crowd pleaser than ever as he sang "There'll Always Be an England," "The Nightingale Sang in Berkeley Square," and "To the Royal Air Force." Guests danced until 1:00 A.M. to the Meyer Davis Orchestra in the main lounge. As bandmaster Davis waved his baton through his final number—"Good Night, Ladies"—guests remaining at this late hour gathered to sing once again "God Save the King," followed by "The Star-Spangled Banner."

Class after class of Riddle Field RAF cadets was overwhelmed by the generosity and true affection that distinguished Palm Beach and its residents when it came to welcoming and sheltering those young flight trainees who had traveled a perilous three-thousand-mile journey to Clewiston, Florida. Florence Nesmith and her equally dedicated "Cottage Colony" associates opened their luxurious, even palatial, homes to the young men who were to spend six months in a very different climate and culture. These were young men who had little concept of what lay ahead.

In her own words, Mrs. Nesmith "had no shame" when it came to cajoling, pleading, or wheedling what we now call "freebies" for "her boys." It didn't take long for the newly arrived cadets to get word from senior class members. The message was simple: "Mrs. Nesmith." Need a place to spend a day or two on a free weekend, but funds are low? Mrs. Nesmith had her handy list of Palm Beach friends who were ready, willing, and more than able to handle that minor detail. And so on with many other favors: free admission to world-class country clubs for the golfers among the cadets, complimentary tickets to theatrical productions, deep discounts at many restaurants, and special rates on rental cars.

And, of course, parties, parties, and more parties. Florence

Nesmith had a magic touch when it came to forging young Anglo boy-American girl friendships. There was a mutual attraction from the start, based in part on "those charming English accents" (even though the accent may have been cultivated in Scotland, Wales, or Ireland). Then there were those outgoing "honey chile," "darlin'," "sweetie pie" girls of Florida, who were quick to clutch the arm of a blue-uniformed RAF cadet and bury her gaze into his eyes. Who could resist?

Many RAF aircrew members visited Mrs. Nesmith after the war. Among them was Peter Spenser, a Riddle Field graduate who had been a guest at the Nesmith home many times during his training days in Riddle Field's Course 16 with Peter Pullan, Peter Orchard, and Monty Manners. The last time that Florence Nesmith saw Peter Spenser in uniform, he had just received his wings and his pilot sergeant stripes. He had only a few days before going back to England for his operational training. There were other RAF cadets on hand at the farewell party and Peter, a fine pianist who often entertained his friends with music at the Clewiston Inn, did not need much urging to play at the Nesmith home. He played until late in the night, while his friends gathered round and sang anything from classical music to the latest hit tunes.

Spenser was assigned to the Transport Command upon completion of Operational Training and flew as copilot on Dakotas, supporting Allied troops in France by delivering supplies. They landed on hastily built landing strips as close to the front lines as possible. After supplies were taken off the aircraft, Transport Command would fly wounded men back to England, where medical teams would be waiting. And then the horrible day of March 27, 1945, just six weeks short of the European war's end, he lay in a hospital bed in England. A crash on the ground left him with his neck broken, his right arm amputated at the shoulder, and his left arm paralyzed. Peter Spenser, after months of hospitalization, started on a slow road toward rehabilitation, and ultimately achieved renown as an artist with the Association of Mouth and Foot Painters. Many of his paintings, executed by holding the brush between his teeth, were exhibited throughout Great Britain to critical acclaim; his fame spread to many parts of the world after news coverage and television appearances.

Peter Spenser was not a "handicapped painter." He was simply a painter of remarkable talent, judged on his abilities alone—abilities that grew over the years. During a trip to the United States in 1966 to appear at several exhibitions of his work and talk about the Association, Peter and his wife, Jane, flew to Miami. There to meet them at the plane were Mr. and Mrs. Nesmith. When the couple walked down from the airplane, Florence Nesmith, then seventy-eight, put her arms around them. "Peter," she said, "you should keep in touch more often. Your room is all ready and you'll have to be here for a good long stay." The Spensers spent several sun-drenched days with the Nesmiths before going on a lecture tour in several American and Canadian cities.

The British government, in 1947, awarded Florence Nesmith His Majesty's Medal for Service in the Cause of Freedom. The British Embassy sent along an apology: "We are currently in short supply of the proper metal. There will be an unfortunate delay in our presenting you with the insignia of your decoration." But the United Kingdom, as always, did come through. A British warship was detailed to deliver the medal in its course of duties in the waters of the United States. Mrs. Nesmith was welcomed aboard the destroyer and presented with the medal by the captain, while officers and crew looked on at attention. "You have many fine boys on this ship," Mrs. Nesmith said. "They remind me of the Air Force boys who used to come to our house." The seemingly indefatigable "mother hen" to hundreds of Royal Air Force cadets during four wartime years kept up her busy pace for more than two decades after World War II. Florence Nesmith played active roles in many community charity efforts, served on the English Speaking Union board of governors for many years, and maintained correspondence with former Riddle Field trainees. Mrs. Nesmith died at ninety-three on February 15, 1976. She was the widow of Ira Nesmith, who died in 1968. Scores of letters extending condolences came from all sections of the British Commonwealth. The Rev. Dr. Roscoe T. Foust of the Bethesda-by-the-Sea Episcopal Church officiated at the service.

"The obituary will state that Florence left no children," Dr. Foust said, "but you need only to look at your atlas or globe to see where so many of her boys are today. Spin that globe and stop it with a finger. England, of course, they are there. Naturally, Scotland and Wales and Ireland; Florence's children are there. They are also in Canada, Chile, South Africa, and Australia. They are in Switzerland, the Netherlands, and the Bahamas. Here in America, they are in several states other than Florida. And there are those who, in the words of English poet Rupert Brooke, 'lie in some foreign field that is forever England.' We may never know how many of Florence's boys lie in those fields made famous by the sensitive Great War poet. But forget them? Never, Florence, never, never."

Chocolates to Pat Taylor

Parker T. Wilson, retired pharmacist, recalls several of the "Wilson family boys" who trained at Riddle and then returned to Great Britain for posting to combat units. Wilson, at eighty-three, recalled some of the cadets who visited them in their Clewiston home or at the pharmacy:

> Pat Taylor always loved chocolate, just couldn't get enough of it, so I sent him a box of milk chocolate for the first Christmas that he was back in England. His mother wrote a thank-you note to me and said that Pat didn't return from a mission over Holland, and that she would let me know if he returned. I never heard any more.
>
> Then some time later I heard that John Stewart-Denham was killed on landing after nine missions. Hugh Williams, a little Welsh lad, flew after the war for BOAC and a boy named Collins became a pharmacist.

In 1943 Wilson wrote a poem entitled "Leaving" that he said reflected his family's thoughts on the cadets that they had known over the Riddle Field years:

> *He stood there talking to us*
> *As he waited for the train;*
> *He was leaving us forever*
> *And would never come again.*
> *He was going home to England*
> *To protect his native land;*
> *He was leaving friends behind him*
> *Who had offered him a hand.*
> *But he knew our hearts went with him*
> *And his youthful eyes were wet,*
> *And we never shall forget him,*
> *Out little RAF cadet.*

Sewing on Proud Wings

Mrs. Beryl Bowden, longtime publisher of the *Clewiston News* and founder of the Clewiston Royal Air Force Museum, remembers well how the subtropical South Florida temperatures made conditions almost intolerable for many English cadets: "They arrived during 90-degree days, still wearing their heavy blue woolen uniforms they were issued over in England in December." Mrs. Bowden was present at several Wings ceremonies, when proud cadets were given their impressive RAF fabric wings. "One by one the cadets were called up to receive the wings they had worked so hard to get," she recalls. "They simply couldn't wait to get them on their uniforms. I sewed several of them on for the boys and knew how proud they were of their wings. I would stuff a bit of cotton underneath to make the emblem a bit more conspicuous."

A 160-mile Movie Trip

Clewiston physician O. F. Schiffli had treated a cadet who had been in an automobile accident for a broken leg, and outfitted him with crutches. Unfortunately, the accident occurred just days before the class of cadets was scheduled to make a bus trip to Miami to see the film *One of Our Aircraft Is Missing,* an English production. The film, starring Godfrey Tearle, Eric Portman, Emrys Jones, and other familiar British actors (Peter Ustinov played a supporting role), centered on RAF pilots who crash-land in the Netherlands and seek to return to England. The bus driver determined that it would be much too difficult for the injured cadet to make the long trip. Knowing that the cadet was bitterly disappointed at being left behind, Schiffli went to check on the boy in sick bay. "Well, son, you don't have any fever," he said after checking the nineteen-year-old's thermometer reading. "You know, I've been thinking that I'd sort of like to see that movie. If I gas up the LaSalle, how would you like to go along?" In about ten minutes the two of them were on their way, heading south and getting ready to chalk up about 160 miles round trip to see the RAF-inspired movie. The injured cadet was elated when they passed the bus taking his fellow students to Miami. When the film ended, the cadets gave it a round of loud applause; so loud, in fact, that Dr. Schiffli, who "wanted so much to see the film," stirred from the nap he had taken during the movie's last half. Really the best part, everyone agreed.

Saturday Night in Arcadia

Saturday night in Arcadia back in the 1940s drew everyone from miles around to the bustling cattle town. Mrs. Burnadetta Johnson has sharp recollections of those days:

> I was in my twenties when the British boys were at Carlstrom. You could always see them at the soda fountains, and any place that sold candy. You just knew that any cadet you saw had a sweet tooth.
>
> I worked at the Plaza Hotel as a desk clerk and remember when a nice English lady managed to come over and visit her son, who was training at the field.
>
> As far as the cadets were concerned, we always found them to be fine young men. All the mothers in Arcadia took them into their homes every weekend. Mrs. Rupert Smith was always making arrangements for the boys to have places to stay as well as lining up entertainment for them.

Smooth Scot Applauded, Cadets "Star" in Film

Cadet James Lindsay Kerr, of Glasgow, was given a standing ovation by members and guests of the Clewiston Kiwanis Club in December of 1942 when he spoke on "Britain at War."

The twenty-two-year-old advanced flight training cadet, son of a Presbyterian minister, was highly praised by the *Clewiston News* in a page-one story devoted to the event. The popular editorial column "Ramblings by B. B." (Beryl Bowden) quoted Kerr as saying "we cadets have exhausted our superlatives in describing Americans and their country." B. B. added: "It is equally true that those who heard Cadet Kerr exhausted *their* store of superlatives about this splendid speaker and his talk. [He is] a speaker with such a clear, concise mind that every expressed thought is as sharp as an etching. James L. Kerr combines the rugged, deeply felt sincerity of Winston Churchill with the poise and charm of Franklin Roosevelt."

Kerr told the rapt audience that he was in church on Sunday, September 3, 1939, at about 11:30 A.M., when word of war was announced by Prime Minister Neville Chamberlain. A parishioner who had been stationed outside the church was monitoring the BBC news in anticipation of such news. The scene could well have been an inspiration for the 1942 film *Mrs. Miniver,* a moving drama about a middle-class English family learning to cope with war, starring Greer Garson and Walter Pidgeon. As the Rev. Mr. Kerr was delivering his ringing sermon, "A Cost of Peace and a Cost of War," the man from outside entered the church, halted in front of the lectern and handed a slip of paper to the Church of Scotland rector. There was complete silence as the minister announced to the congregation: "Dear friends, since eleven o'clock this morning our country has been at war with Germany. You will show no lights after dark and you will carry your gas masks with you at all times."

Cadet Kerr contrasted his father's solemn message, with its implication of burdensome toil and self-denial, with the adventurous spirit at the outbreak of the World War in 1914. "That was the war that we entered into with buoyant, adventurous spirit," Kerr said, "a sort of Sir Galahad in search of the Holy Grail." But the British had learned the hard lesson of the Great War, Kerr said. More than 909,000 young Britons had died in the four-year struggle, and there were 2,000,000 who suffered wounds. "The suffering of World War I had taught us that war is the dirtiest of dirty tasks," Kerr said. "We are in the Second World War with no dreams of glory, only the grim reality and determination to fulfill an awful, inescapable duty."

The poised, articulate speaker spoke of Britain's "darkest hour," the 1940 evacuation of British Expeditionary Force troops off the Dunkirk beaches in northern France. "We had lost all of our equipment and many of our best men lay in France," Kerr said. "But acting as a prophet of dawn was the Royal Air Force, a puny fledgling, bitterly outnumbered on every count, and yet managing to keep Hitler's forces away from the British Isles." Kerr concluded by saying that Britons are convinced that Americans are "the most friendly people on earth." He said that many RAF pilots trained in the United States are "sure to carry a wonderful message of good will when they return to their home country."

A large contingent of Sarasota residents, headed by David Lindsay, editor of the *Sarasota Herald-Tribune,* joined with Arcadians to honor RAF cadets of Carlstrom Field on November 15, 1941. The young

British men were invited to a special "March of Time" feature preview at Arcadia's Star Theater. The Saturday morning unedited presentation on the training of British Royal Air Force cadets in the United States was produced by New York motion picture crews. The theater was filled with city and county officials and local citizens who came to see the cadets in movie action. The film showed the students doing various flying exercises, including tight formation patterns. Also shown in detail were classroom activities, the mess hall, the dormitories, and the athletic field as well as excellent footage of beaming cadets cavorting on the beach at Sarasota, at Lido Beach Casino, enjoying the Ringling Museum, gaping at residents of the Reptile Farm, and finally being driven back to base by Arcadians and Sarasotans. As cadets recognized fellow students, they would shout out appropriate (primarily) remarks, mostly consisting of "inside jokes" concerning girlfriends, flying goof-ups, personal appearance (oh, those dainty legs), and other such ways to "take the mickey out of" (make fun of) friends. When the feature ended and the house lights went up, the cadets gave the picture a rousing cheer. Comments to hosts were highly favorable. Cadet Richard Williams, of Swansea, Wales, displayed a touch for newly acquired American slang: "Oh, we got a big kick out of it."

Rodeo Town Lassoes Cadets

Arcadians staged the most festive reception within memory of the DeSoto County rodeo town on June 12, 1941. It was *the* big day in Arcadia, the day that some thought would never come, the day that the British were coming to energize Carlstrom Field for training of Royal Air Force pilots. The reception committee wasn't fooling around. They knew just how to arrange those tables placed on the Arcadia House lawn and had met days in advance. Sharing the planning duties were Miss Ruth McElya, Mrs. George Stonebraker, Mrs. Howard Shaver, and Mrs. Rupert Smith. Miss Emma Marie Vance, DeSoto County Cowgirl Queen, had her troupe of local cowgirls on hand to welcome the first quota of British cadet-pilots. Tables were piled high with pyramids of oranges. There were gallons of chilled orange juice, with rivulets of moisture streaking down the frosty, fat jugs. Freshly brewed coffee shared the platform with hot tea—the British preference.

Mrs. Stonebraker was sure that the British cadets wouldn't want iced tea. "They don't even like cold beer," she said.

"And how do you know?" asked Mrs. Shaver, in an offhand, teasing manner.

"I read it in *Liberty* magazine," said Mrs. Stonebraker. "And they don't like soft drinks."*

Sarasota Leader Drives Cadets

A prominent Sarasota businessman was a willing chauffeur for British cadets who enjoyed coming over to the Gulf Coast city for long weekends, usually Thursday evening until Monday morning. Frank Berlin, in his mid-thirties during the early war years, was born in Kankakee, Illinois, on the river that shares its name, about one hundred miles southwest of Chicago. Berlin's family moved to New Mexico when he was two because his mother suffered from respiratory problems and was given six months to live unless she had a change of climate. The doctors had it right: she lived to be ninety-five. Berlin spent his boyhood punching cattle and working in a general store that his parents owned. He was sent to Chicago, where he studied pharmacy and later became associated with the large Walgreen Drug chain. In the early 1930s, Berlin traveled the southeast, convincing investors to open a Walgreen franchise. His sales presentation was so good that he talked himself into opening one in 1938, Bradenton's first Walgreen store. His Walgreen outlet in Sarasota opened in 1940. At age eighty-eight, the highly successful drugstore operator, real estate magnate, advertising agency executive, and community leader looked back on the days when blue RAF uniforms were commonplace along Sarasota's Main Street:

> The boys enjoyed Sarasota in those days. It was a place that offered a great number of places to go and the people of Sarasota really made them feel welcome. It wasn't as though they were a bunch of strangers.
> Many people had the students as guests, and I tell you that they never had to walk far when they wanted to go someplace. I used to

*She was right on one count: iced tea was never a big hit with the cadets. But Coca-Cola became the beverage of choice for many of them.

take them back to Arcadia when they needed lifts. It was interesting to just drive along and listen to the cadets tell of the good times that they had as they talked among themselves.

Frank Berlin glanced over the well-appointed office he still maintains, looking over the framed proclamations, autographed photographs of well-known people, and newspaper tributes to his business triumphs and community activities. "It's great to look back on all those years," he said, "and I really must say that I got as much enjoyment in listening to those RAF fliers as anything that I've ever done. They were a great bunch and I often wonder what happened to them after they went home. It's too bad that so many of them were casualties in that terrible war."

Cricket Made Simple

Frank Rollins, a former Bradenton technical writer, will never forget a description of the game of cricket he heard in 1943 over a glass or two of beer in an RAF hangout bar.

What happened was that a guy from Georgia was yakking with a British cadet and they got to talking about sports.

Both of them had a couple too many, and the RAF flier said he thought American football was a "bloody awful farce with rules that don't make much sense."

Then the Yank said that anyone who thought that football didn't make sense didn't know diddly. Cricket was the most stupid game there was when it came to dumb rules.

The argument over the confusion factor of football rules versus those of cricket drew to a merciful end when the British cadet said, "Cricket is a simple game. Let me sum up the rules for you in simple terms," he said as he finished off his glass.

"In cricket you have two sides, one out in the field and one in. Each man that's in goes out and when he's out he comes in and the next man goes in until he's out.

"When they are all out, the side that's out comes in and the side that's been in goes out and tries to get those coming in out.

"When both sides have been in and out, including the not outs, that's the end of the game. Do you understand?"

"Hell yes," said the Yank. "Makes sense to me, buddy. Let's have one for the road."

RAF Wing Commander Harry Hogan made an inspection at Riddle Field one day and noticed a group of U.S. cadets learning the fine points of soccer at the hands of British cadets. Hogan asked one of the Yanks if he understood the difference between soccer, rugby, and hurling, Ireland's national pastime. An extremely rugged game similar to field hockey, hurling is played with a cork-filled rubber ball covered with horsehide and a "hurley," a curved, broad-bladed wooden stick.

"No sir, I don't know the difference," replied the American as he stood at attention.

"Well," Hogan said with tongue firmly planted in cheek, "soccer is a gentleman's game played by hooligans; rugby is a hooligan's game played by gentlemen, and hurling is a hooligan's game played by hooligans."

"Mamma" of the Everglades

Remembered with fondness and gratitude by many Riddle Field cadets is the Wadlow family of Palmdale, Florida, hard by the Everglades. Carolyn Wadlow, "Mamma" to hundreds of cadets, always had her friendly little house open to any of the cadets who wished to enjoy a few days of rural Florida life. A notation in an edition of the 5BFTS Association newsletter carried a tribute to Mrs. Wadlow:

> "Mamma Wadlow" was always generous with gifts and motherly affection for her youthful "flying boys." Her son, Ralph, enjoyed showing off his beehives and imparting his intimate knowledge of Seminole Indians, alligators, and other tidbits of Florida lore.

Mrs. Wadlow was not a "hug and farewell" person when "her boys" left Riddle Field but kept in touch with them as they were assigned to distant posts. They, in turn, wrote to her out of love and affection. Embry-Riddle's publication, *Fly Paper,* carried many of her letters to the editor as she reported on goings and comings of Riddle graduates who had shared with her a bonding of spirit and love of freedom. Excerpts from her letters follow:

> Douglas Pollard, Course 9, is reported missing. He was on his seventy-ninth mission and would have finished "ops" (operational duty) in a few weeks.

D. D. Campbell, also Course 9, who instructed in Texas and later in Scotland, is on operations as a bomber pilot.

Ralph Mullins, Course 16, transferred to the Navy and is now flying carrier-based Spitfires.

Peter Pullan, Course 16, has been posted to Burma after being wounded in Holland and returning to England.

Mike Hills, outstanding cadet of Course 14, is instructing on primary trainers in England.

Peter Taylor, another top cadet of the same Course, is instructing on twin-engine aircraft.

Fred Cox, also of Course 14, is now on gliders in India. He has a "mere" 1,000 hours and is due for a rest.

Arthur Wyman, Course 18, is in active service in India, as is Harold Prust, Course 9, who went to engineering.

Ivan Harper, Course 18, is a navigational instructor and J. W. L. Ivimy has been released from service to return to his essential work as a civilian.

Alex Whittle, Course 9, was a prisoner-of-war and word has it that he has been released.

Douglas Coombs, Finlay McRae, and Alan Head, all of Course 19, are training on gliders. Paul Jackson, of the same Course, is doing staff work, but gave no details.

In her report to the *Fly Paper*, Carolyn Wadlow enclosed a birth announcement card and a newspaper clipping. The birth announcement was of Christopher Harold Marshall, born on March 30, 1945, to Pilot Officer and Mrs. W. H. Marshall. The news clipping contained a notice of Pilot Officer Marshall's death during a bombing operation over Germany on February 24, 1945.

Legendary Flier Awes Youth

Tom Keller was a ten-year-old Miami schoolboy in 1973 when his father took him to Riddle Field in hopes of meeting some former RAF cadets, who were having their thirty-year reunion at the field where they learned to fly. Tom was an avid builder of model airplanes and had won a Boy Scout merit badge for his construction of a Spitfire, his favorite fighter plane.

"We were hoping that maybe some of the British veteran pilots might do a little flying," Keller said, "but the old field wasn't in operation and, besides, it probably would have been against FAA regulations." But young Keller did see many of his hopes fulfilled. He took along his model Spitfire and had it autographed by Ian (Jock) Blue, a legendary former RAF pilot who had won his wings at Riddle. Blue told Tom Keller about his 200 operations over enemy territory and how his aircraft had been damaged by ground fire on 13 successive missions. The stocky pilot took Tom over to an old AT-6 advanced trainer and lifted him into the cockpit. "This is what we flew down here, Tom, when we were in our final training phase," Blue told the wide-eyed lad.

More than twenty years after that visit to Riddle Field, Keller still cherishes memories of that trip to Clewiston and the friendly treatment he received. "I sure had a lot to tell the other kids in fifth grade," he said, "and I still have the Spitfire. We even got to see someone fly over the field in an AT-6, a real thrill for me." (Piloting the AT-6 that day was Bob Bennett, son of the former Riddle Field instructor Marty Bennett.)

"Girls" of Riddle Field Look Back

Mrs. Lewis H. Blount of Clewiston recalls those exciting days in the mid-forties at Riddle Field. She then was sixteen-year-old Lois Heflin, recently graduated from high school, and starting on her first job as a Riddle Field secretary. "I worked for Riddle-McKay Aero College at Riddle Field from May, 1943 to August, 1945," Mrs. Blount said when interviewed. "First I was in the personnel department and by the time the war was over, I was secretary to General Manager E. J. Smith and James Durden, assistant manager."

Debbie Hatfield of the *Clewiston News* wrote a feature story about Lois Heflin Blount that ran February 11, 1981. The feature read in part:

Today Lois Blount recalls a lot of old memories about the old air field, Airglades as it is called today.

The aero college, which taught British RAF cadets to fly during World War II, employed many persons from Moore Haven and Clewiston, Mrs. Blount said. "My salary was seventy-five dollars a month, not bad for someone fresh out of high school in those days."

The story went on to relate the reactions of the British cadets to south Florida customs. "They were used to strict rationing back home," she said. "If you would take a tea bag out of a cup, they would be shocked. They would say that you could always use that tea bag again."

The RAF cadets dated the local girls, much to the irritation of the young men of Clewiston, who called the cadets "those darned limeys." Lois dated a Scottish cadet from Edinburgh. The couple would have supper at the Riddle Field canteen, take a bus to Clewiston, and see a movie for twenty-five cents. She recalled a favorite meeting place for cadets and their dates. It was the Canteen Club operated by Mrs. W. C. Owen, complete with jukebox and all the trimmings. Best of all, no charge. Even the jukebox didn't take money.

On graduation day, when cadets received their coveted wings, the canteen would be packed with graduates who would pile into their favorite spot. Then they would gorge themselves on ice cream and milk shakes in a big splurge. It was their last chance because soon they would be off to England, where ice cream was a memory of things past.

Tempe Jean Garlick, of Lehigh Acres, Florida, was a high school classmate of Lois Heflin (Blount), and worked at Riddle Field in the timekeeper's office during the last year that the base was in operation. She recalls:

I met my husband-to-be, an RAF cadet, on December 13, 1944, in Clewiston, when he and two of his cadet buddies were hitching a ride to the field. There would be a string of two or three cadets waiting for rides.

To be hospitable, I invited the cadets who had been given a ride to stop by some time and use the game room that my landlady had set up in the apartment building. I couldn't place their faces because the interior light in my father's car wasn't working.

When I went to my apartment building the next day I didn't recognize some RAF cadets who were in the recreation room, but learned they were the very boys I had invited.

It was then that I was officially introduced by my gracious landlady to the man who was to become my husband. I had no idea that he was one of the cadets who had hitched a ride in our car.

John was twenty years old when he graduated in Course 23 with his wings. The war was over in Europe, and it was a blessing to all that it was finished in Japan soon after. He never saw any combat service.

We soon became engaged to be married and we spent all of his free time from early 1945 to June seeing each other. John spent all his weekends at my family's house located between Moore Haven and Clewiston.

During the year we spent apart when he was returned to England, our postage costs were enormous. I remember receiving an envelope from John that contained a letter and a linen hankie—a rather lumpy linen hankie in which he had sewn an engagement ring. In the British style, the ring was rather flat, more like American wedding rings. I was surprised that it came through the mail service with no problems.

I went to England the first of July, 1946, and we were married a few weeks later in an Anglican Church. We stayed in England until August, 1947, preparing for our return to the United States. At that time, people were still getting transported by troop ships. We managed to get our names listed for passage to New York. John shared space in quarters with 125 other men and I was in quarters with seventeen women.

I must say that I enjoyed my stay in England so very much. My in-laws not only accepted me warmly, but made me comfortable in their small flat. Their home had been destroyed in the air raid while they, as air wardens, were away as bombs fell on their house. They were alerting others to German air bombardment dangers.

My "English family" could not have been better. I enjoyed England so much. My mother-in-law was full of life, interested in everything. I spent many hours in Kew Gardens, cycling with "Mom" and "Auntie Dorothy" to Windsor Castle; and then taking the train to Dover, the white cliffs.

When spring arrived after the coldest winter in fifty years, I stood in awe of the most magnificent bounty of flowering trees. I felt it was *my* England.

The English people were wonderful; I enjoyed so much getting acquainted with all my husband's relatives and friends. I lost John due to cancer after a good and happy marriage of thirty-one years. My last trip to England was in 1989.

Grads of '42 Worries

Sarasota High School graduates, class of 1942, were a special group. The graduating seniors who marched to "Pomp and Circumstance" on that special June night were the first class of eighteen-year-olds since Pearl Harbor. World War II was raging, unemployment was only a word, and they would have no trouble finding a job.

Sarasota Herald-Tribune writer Robert King wrote a fifty-year retrospective on the sometimes fortunate, but often star-crossed, members of the Class of '42. The school had promised diplomas to boys who left early to join the service. Twelve seniors took up that offer and reported to the recruiting office. Two of them were killed in action before the class graduated. Valedictorian Laura DeWalt Queen was quoted in the King article of May 31, 1992: "It was a very bleak future and the whole world was so unsettled." She recalled, however, declaring in her address that "we are not afraid," an attempt to hide the uncertainties sweeping the 109-member class. Royal Air Force cadets from Arcadia frequently stayed with Sarasota families and dated local girls, who, Laura Queen said, were "agog over the wonderful-looking Englishmen." Bradenton, Sarasota's neighbor to the north, was another favorite visiting spot for Carlstrom Field cadets. The Manatee County seat welcomed the British cadets to mixers sponsored by service clubs and churches, where the local girls usually outnumbered the cadets by two to one. Mary Stuart, of Columbia, South Carolina, lived in Bradenton during the war years, and with her friends sought to impress the British cadets:

It was great for the boys. The cadets had their pick of the lot of girls. So many American boys were in the service that it created a real shortage of dates. We always enjoyed listening to their accents and they liked to hear our southern way of speaking.

To tell the truth, we would exaggerate our southern accents so much that we all sounded like the cast of *Gone With the Wind.* You know, saying things like "I do declare, honey," "if that isn't the livin' end," and then throwing in a bunch of "you-alls" for effect. I got into the act so much that I couldn't turn off the Scarlett O'Hara deep south corn pone routine at home. I was talking such phony southern that I was making "is," "and," and "but" sound like two-syllable words. Sort of like "ayyund," "eeiss," "buaht."

I'll never forget the day I was talking on the phone to a girlfriend while my father was reading the newspaper. When I hung up, he put aside the paper and quietly said in a fake molasses, corn bread, and grits accent, "Now come on heah, Miss Melanie, you've got your old daddy's head a spinnin', honey chile." Reverting to his smooth south

Florida speaking habits, he said, "In other words, kid, it's about time to start talking like a normal person." And that's what I did, starting right then. I left Tara behind me forever. Fiddle-de-de.

Hotel Sebring Is Best-Kept Secret (Hotelier Higgins Praised)

Hundreds of British RAF cadets who took that long, tiring train ride from Moncton, New Brunswick, Canada, to Clewiston, Florida, between 1941 and 1945 have vivid recollections of Sebring, Florida. It was there that the young men on their way for flying training at Riddle Field made a stop that men in the service usually only dream about. After looking out sooty train windows for hundreds of miles at nothing but flat cattle-country grazing land, stretches of cloned orange groves, and dusty one-mule-and-a-wagon "cracker towns," they weren't prepared for the spectacular welcome. A line of just-polished cars, many driven by attractive young ladies, were waiting to transport them from the Seaboard Railroad Station to the four-story Hotel Sebring, one of the most attractive buildings in the charming little city among the rolling lands of Highlands County.

The RAF men were greeted at the door by an impressive-looking man in his forties with a ready smile and military bearing. He was William Vernon Higgins, owner-manager of the sparkling hotel on South Palm Street. Vernon, as he chose to be known as, was the son of Edward John Higgins, the first elected general of the Salvation Army. Vernon had been a pilot in the British Royal Air Force. Higgins always looked forward to seeing the cadets as they made periodic stops at Sebring before heading down to Clewiston on the last leg of their trip from Canada. He sometimes would arrange for Sebring area girls to dress in Spanish costumes and serve specially prepared meals of regional favorites to his young guests, who were only too glad to obey his command: "Eat hearty, lads! There's more where that came from." Homemade desserts followed several choices of entrees and, to top off the lavish meal, sacks of oranges were delivered to the railroad carriages so that the cadets could have multiple tastes of Florida on the way down to Riddle Field. Each cadet was given a supply of cigarettes and sweets plus picture post cards of Sebring to send home.

On March 15, 1945, Higgins received a letter from Captain Frank Fernihough, former commander of the 31st Royal Air Force Depot in

Canada, commending the hotel man and the Sebring community for the many kind gestures during the war years. Captain Fernihough wrote:

> Now that the Royal Air Force Depot has disbanded, I am returning to the United Kingdom. I should like to express to you my appreciation for the many kindnesses that you have shown the aircrew trainees en route from this depot to Clewiston.
>
> There is no doubt that the training of these young pilots in such admirable conditions in Florida was proved an excellent thing, and your initial hospitality that has been shown to them is sincerely appreciated.

Because of the many glowing comments made by former RAF cadets about Higgins and his fabulous hotel, one might think that a generous amount of source material would be available. Interestingly enough, this is not the case. Our basic documentation came from RAF veterans living in England, Scotland, South Africa, and Canada. The Sebring Historical Society made available some old bound volumes of now defunct newspapers and one small clip helped round out the story, thanks to Joan McAfee, president, and Lois Thiele, archives researcher. The clipping stated that the hotel had been a stopping-off point for RAF cadets coming to Florida from Canada and quoted the praise from Captain Fernihough.

Cadet Returns to Sarasota

The friendship and hospitality of the John Levinson family of Sarasota was so warm that RAF cadet Berkley Barron (see Barron, p. 45) returned to the place that he had enjoyed so much as a Carlstrom cadet. Eleene Levinson Cohen and her brother, "Bud," were with a group of young friends at the Lido Beach Casino in early 1942 when they saw a small group of RAF cadets who looked a bit lonely despite festive activities going on all around them. "We decided that it was the least we could do to invite them over," Eleene recalled, "and that led to friendships that have lasted for all these years. The boys told us they were looking for a place to stay for the weekend. Bud said, 'You have a place to stay,' and gave them the key to our family's Siesta Beach cottage. From that time on, the boys came to the cottage at every opportunity, whenever they could get free time. After they left Arcadia for other training fields, they kept in touch with us by mail. We are deeply saddened to learn that one of the boys was killed on his first mission. Berkley, of course, moved to Sarasota and became a newspaper executive."

The Levinson family members were owners of the women's ready-to-wear clothing store, The Sport Shop, a Main Street fixture for sixty-one years. Eleene's husband, David Cohen, was in the U.S. Army and stationed in the South Pacific during the war. He later served as a Sarasota city commissioner and mayor and was a driving force behind establishment of Sarasota's nationally acclaimed New College.

Arcadia Restaurant Packs in Cadets

If you visit , don't miss Wheeler's Goody Cafe (and don't forget to order the daily pie special). The cozy restaurant has kept locals and visitors coming back for sixty years. In the early 1940s the Goody Cafe served up tasty food around the clock. No opening or closing hours. It was a favorite hangout for the RAF cadets at Carlstrom Field.

"My goodness, those boys sure had big appetites," eighty-two-year-old Alice Wheeler said. (She takes pride in being the oldest business person in Arcadia.) "Sometimes there would be so many of the British cadets that we would have to close the door until some of them left," she said. "Then one of them would open the door and let a few of his buddies in. It got to be a real hassle, but they were really good kids. Oh, the cadets always had something to do in Arcadia. We had a busy USO Club and the town was always trying to make the boys feel at home. I know that some of them were homesick, though, even if they tried not to show it." Mrs. Wheeler ran the restaurant with her husband, C. B. Fiegel, from the mid-thirties until his death in 1951. She later remarried and the restaurant then took on the Wheeler name. She has twin children from her first marriage, Jerry Fiegel and Jeanette Fogel, who live in Arcadia and say they wouldn't live anyplace else. "I remember that Jerry and I used to stand out in front of the USO in the evenings," Jeanette said. "While the music was blaring out the forties stuff—Glen Miller and all that—we would start singing the 'Wild Blue Yonder' or 'The Caissons Go Rolling Along.' And we could sing loud, too. Just a couple of pranksters. But that's when our mother took charge and sent us to our rooms."

Arcadia was a spot favored not only by British cadets; the DeSoto County seat was part of a select group. Author Norman Clampton in 1992 picked Arcadia (population 6,488) for inclusion in his book *The 100 Best Small Towns in America* (Prentice Hall General Reference). "Oh, shucks," said Mrs. Wheeler, "we could have told him that a long time ago."

In 1994, Mrs. Wheeler died and her landmark café was bought by Eddie Tang, a popular immigrant from Hong Kong. Tang has kept the pie menu intact: Monday and Friday, chocolate and chocolate coconut; Tuesday and Thursday, lemon; and Wednesday, butterscotch and butterscotch peanut butter.

Sarasota's "Brooksie" Bergen Tells of RAF Friends

Multitalented Bernice Brooks Bergen is an author, a freelance newspaper columnist, an artist whose paintings and watercolors are in private collections throughout the country, and an accomplished actress. She has appeared in many plays and musicals in professional theaters. The spirited Bernice (Brooksie) was an active member of the young set that made Sarasota a leading fun spot during the 1940s. Brooksie and the rest of the college circle were regulars at the Sarasota night spots that flourished on the wave of a wartime economy. Mingling with civilians in the crowded, noisy bistros were men (and some women) in uniform who were stationed throughout the Gulf Coast area: Sarasota, Venice, Fort Myers, Sebring, and other bases. Royal Air Force cadets from Arcadia's Carlstrom Field and Clewiston's Riddle Field cut dashing figures as they, too, jammed elbow to elbow in the dimly lit rooms that were clouded with thick nonfiltered cigarette smoke.

Brooksie Bergen's 1993 book, *Sarasota Times Past* (Valiant Press), is dedicated to her husband, John Bergen, for "love and support." In writing her retrospective of early Sarasota days, she recalls the attraction that Sarasota held for servicemen:

> Whenever the young men had passes, they converged in Sarasota because word got around that some of the prettiest young girls in the

area lived there. It was also one of the more populated towns in the state, hosting the Ringling Brothers and Barnum & Bailey Circus winter quarters and spring training for the Boston Red Sox. The media were enthralled with the fabled Ringling Museum, famous personalities, and baseball greats who wintered in Sarasota.

The girls in those days felt it was important to look pretty and feminine for the boys in uniforms. Hair was long and flowing. Bare tanned legs were shown to advantage by abbreviated skirts since silk stockings were in short supply and even leather shoes were rationed.

There were a great many shortages, but no shortage of love and patriotic spirit. Sarasotans would often drive around Five Points, inviting every lonesome-looking soldier to join the family for dinner.

When interviewed, Brooksie Brooks Bergen gave this account of how Sarasota families went all out to make the RAF cadets at Arcadia feel welcome at regular special events:

The RAF boys were brought by bus to Sarasota for dances that were heavily chaperoned. Then there were times we were bused to Arcadia, accompanied by hard-eyed chaperones ready to stamp out even the slightest hint of hanky-panky.

I met Syd Tucker, an RAF cadet at Arcadia, at one of these dances. He was a charming, articulate, and rather cocky airman who washed out of pilot training and then trained to be a gunner on a bomber. Syd spent many hours at our home, having dinner with us and fishing off the pier. He returned to England and was shot down during a mission in North Africa. His mother continued to correspond with us for several years. It is sad to recall that most of the RAF cadets that we met were killed during their operations.

All the British cadets we knew had wonderful manners, together with a great love of country and home. I remember that my sister and I met some marvelously attractive Scottish cadets, one of whom was Gilbert "Gibley" Stuart, who became my sister Jinx's boyfriend. The cadets gave us their wings and I will never forget them.

I was very smitten with Michael Sullivan, who wrote poems to me after he returned to England. His letters stopped all at once and I never found out what happened to him. I remember so vividly that the British cadets loved to sing. They sang "The Bells of War Go Ring-a-Ling-a-Ling," "I've Got Six Pence" and other songs. These were very special young men in those very special days.

Charlotte Steele Anderson and Her Dad's Great Tavern

Charlie Steele was all set for the RAF cadets from Carlstrom Field when they showed up at his Punta Gorda tavern on weekends. It was simple, really. Charlie merely loaded up platter after platter with deviled crab, fried mullet, and anything else that captured his fancy. "These boys can really pack it away," Charlie always said to other patrons. His popular good-food, good-drink, fair-price establishment was a magnet to persons living in the then sleepy fishing town on Charlotte Harbor. "My father had a large Dutch kitchen and dining room in the tavern," said Charlotte Steele Anderson during an interview. "I was too young to really appreciate everything about those days. I do remember that the RAF boys came over many times and that Dad did always see to it that they had plenty to eat without it costing them a lot of money. One of the boys gave me an RAF emblem."

A newspaper story reported that there was a piano and a British-born lady named Melody, married to a Punta Gorda store owner, who sang with the cadets. "Knowing my dad, that probably happened," said Mrs. Anderson, "but I don't recall anything like that."

Yanks Taught 'Em to Fly, but Singing Came Naturally

Sylvia Stanton and Gloria Hansen were part-time waitresses at the Hotel Sebring and called in for banquets or large dinner meetings of business groups. They were usually available on those days that the southbound train transporting RAF cadets from Canada to Clewiston stopped at Sebring for a few hours. Tired cadets were driven in a caravan of cars to the hotel, were given a chance to freshen up, and then sat down to a "smashing" breakfast or lunch depending on the time of arrival. There would be music along with the fabulous meal, adding such a festive tone that the British cadets would often burst into song, much to the delight of their Sebring hosts. "They were really great singers," Sylvia said. "I mean they knew the music and the words." "We would stand in the back and lead the applause," recalled Gloria. "Everyone in town was surprised at how they sang as they went back to the train station." You could bet on it. Whenever a group of British cadets got together, it didn't take long for them to

234 :~ *Will Largent*

start singing. Two Americans were just that, two Americans. Two Britons were a duo, three, a trio, and four, a quartet. Throw in a Welshman and you would have enough talent and power to drown out the Mormon Tabernacle Choir. When British cadets rode on trains or buses, they sang; when they were together at a railroad stop waiting for a train, they sang. They sang with gusto the old songs instantly recognized by anyone within hearing distance.

British servicemen, unlike Americans, enjoyed group songfests, much to the surprise, occasional derision, but frequent delight of the U.S. forces. "I Was Born in Glasgow" brought cheers and showers of cigarette packs and dollar bills from New Yorkers when Clewiston-bound cadets waited for trains at Grand Central Station. A dozen or so Carlstrom Field cadets (Anglicans all except for an Irish tenor) kept the Arcadia First Methodist church congregation enthralled with "Amazing Grace." Then, to prove their ability to switch denominations, the fresh-faced cadets gave a rendition of "The Old Rugged Cross" that brought tears to southern eyes. When the service ended, the cadets were given hearty embraces by the women of the church, firm handshakes by the men, and sidelong glances by the girls. Invitations to dinner kept the British lads busy throughout primary training before they were sent away to basic. And sometimes at base camp the British cadets would sing songs containing bastardized lyrics of cherished tunes that caused uptight American officers to go bonkers. A popular RAF number that was sung by classes at Clewiston reflected the age-old browned-off (British), pissed-off (American) attitudes toward higher military authority, a sentiment of fighting men of all uniforms dating back to the first time one soldier was told to take orders from another. It is sung to the tune of "John Brown's Body."

The Firth of Flaming Forth

We had been flying all day long
At a hundred flaming feet,
The weather flaming awful,
Flaming rain and flaming sleet,
The compass it was swinging
Flaming South and Flaming North
But we made a flaming landfall
In the Firth of Flaming Forth.

Ain't the Air Force flaming awful?
We made a flaming landfall
In the Firth of Flaming Forth.

> *We flew the North Atlantic*
> *Till it made us flaming weep,*
> *The sea was flaming wet,*
> *And flaming cold and flaming deep.*
> *Operations Room at Thirty Wing*
> *Is simply flaming rotten,*
> *And Two-Six-Nine will be there*
> *Till they're flaming well forgotten.*
> *Ain't the Air Force flaming awful?*
> *Two-Six-Nine will be there*
> *Till they're flaming well forgotten.*

> *We joined the flaming Air Force*
> *'Cos we thought it flaming right.*
> *But we do not care if we fly*
> *O if we flaming fight.*
> *But what we do object to*
> *Are those flaming Ops Room twots*
> *Who sit there sewing stripes on*
> *At the rate of flaming knots.*

The Firth of Forth is an estuary of the Forth River in southeastern Scotland, flowing into the North Sea. Details are not clear as to why "landfall" was made in the Firth. The word "flaming" is a euphemism for a much more colorful, blunt Anglo-Saxon adjective that may come to mind.

L'Envoi: The Terrible Cost of War

. . . What of the aircrew, the fliers, the ones who left their burnt bones scattered over all of Europe? In those young men we may discern the many faces of courage, the constitution of heroes: in lonely cockpits at dizzy altitudes, quartering the treacherous and limitless sea, searching the desert's hostile glare, brushing the peaks of high mountains, in the ferocity of the low-level attack or the long, tense haul of a bombing mission, in fog, in deadly cold, in storm . . . on fire . . . in a prison camp . . . in a skingrafting hospital. For them there is no prouder place, none deserving more honor, than the Right of the Line [place of honor in a military formation].
John Terraine, *A Time for Courage*

From the start of the war (1939) until the end (1945) there were almost 71,000 Royal Air Force fliers killed or reported missing in action, with the majority of them (68 percent) represented by 48,000 Bomber Command aircrew. In addition, there were 8,400 bomber aircrew killed in training accidents not related to combat operations and 1,600 who were killed or died from other causes.

In the summer and autumn days of 1940, it seemed there was a fair chance that even the indomitable spirit of the British would not be enough to stave off an invasion by Germany. The Fighter Command rose to the challenge—and ever so gloriously. The Battle of Britain, and the slim victory over forces superior in strength, gave the British "breathing room," and made it possible for the Bomber Command to hammer assaults on Germany, exacting such tremendous cost in loss of lives, homes, industry, and spirit. It was supposed to be the prelude to the invasion of England. That was the dream—and promise—of Hitler, in early August of 1940, after France had fallen and the RAF strength could be tallied up with simple arithmetic.

It is generally accepted that the battle started on August 13, 1940, and lasted until October 31, a two-and-one-half-month struggle of wills. German bombers attacked coastal defenses, radar stations, and shipping, then shifted in late August to RAF installations and aircraft factories in an effort to gain control in southern England. Although heavily outnumbered, the RAF put up a defense that called upon all the reserve stamina that pilots, ground crews, and support personnel could muster, with the result that Germany gradually gave up hope of ever mounting an invasion of England. Germany had suffered its first major failure of the war.

From early August until October 31, 1940, the estimated loss of British aircraft in the Battle of Britain hovers between figures of 915 to 1,017 for Fighter Command aircraft and 118 for Bomber Command aircraft, as well as 130 Coastal Command aircraft. Total aircraft lost, using top estimates: 1,265. Killed were 537 Fighter Command pilots, 718 Bomber Command aircrew, and 280 Coastal Command aircrew, for a total of 1,535.

German aircraft losses were reported by their command headquarters at about 1,800 aircraft and combat deaths of 2,662 aircrew. Famed RAF Squadron Leader Douglas Bader questioned the reliability of the German aircraft and casualty report. "My own view is that we will only know the correct number of German casualties when the English Channel and the Thames Estuary are pumped out," he said.

Staggering though the losses were, it is generally accepted by military historians that the casualties would have been infinitely greater and the war would have lasted much longer if it had not been for the "scorched earth" policy of Air Marshal Arthur T. Harris. "Bomber" Harris was appointed Commander-in-Chief of Bomber Command in February, 1942, and wasted no time in striking the Ruhr Valley in April of that year. He gave Germany a sample of what lay in store for them by hitting strongholds in the Ruhr, including Essen, with more than 300 bombers. Then, with only three months behind him as Bomber Command Chief, Harris rocked the German high command by throwing 1,000 aircraft against Cologne, a move that put Britain into a strong offensive position. Germany took on the unwilling role of defensive player.

Known as the "Thousand Plan," for the number of RAF bombers flown in attack, the bold May 30 raid was heralded as a turning point in the war. More than 900 of the 1,000 aircraft reported at post-operation briefings that they had hit their targets, with a total of almost 1,600 tons of bombs, high explosives, and incendiaries combined. By any standard, the raid was a thumbs-up success. Almost 250,000 persons were evacuated from Cologne and the surrounding area following the historic RAF raid. More than 260 factories were destroyed; water supplies and other facilities were disrupted. Hospitals were operating on emergency basis only. More than 300 acres of Cologne's city center were brought to ruins. No considerable part of the city was without damage. More than 3,000 homes were destroyed, almost 10,000 were damaged, and 45,000 Cologne residents were rendered homeless for a period of time. It was fortunate that the Cologne raid resulted in fewer civilian deaths than were first expected, somewhere between 1,000 and 6,000. The actual total was closer to 600. There were several thousand civilians wounded, many of whom later died.

Harris held the Bomber Command post from February, 1942 through 1945. He was showered with adulation from those who supported his policy of an all-out offensive position, and with strong criticism, bordering on vilification, from those who claimed there was no moral justification for the raids he ordered on Hamburg, Dresden, and other German cities. Harris wrote in 1942, when advocating the use of high-explosive bombs: "What we want to do in addition to the horrors of fire is to bring the masonry crashing down on the Boche, to kill Boche and to terrify Boche." Harris had a simple philosophy when it came to war: carry the battle to the enemy. Destroy their cities as they would destroy ours. Destroy their homes before they destroy ours. Then go back and do it again. (An estimated 300,000 German civilians died as a result of Allied action during the war.)

There is no better example of the devastation rained upon German targets than the Battle of Hamburg. When Harris spoke of terrifying the enemy, his 1943 summer attack on the Elbe River seaport could have been a blueprint for horror. Bomber Command attacked Hamburg on the nights of July 24, 27, and 29 and August 2, with more than 2,450 aircraft dropping bombs on the city. The RAF massive bombing was bolstered by daylight bombing on July 25 and 26 by about 400 B-17s of the United States Eighth Air Force. Estimates of civilian deaths due to the late July Hamburg attack range from forty thousand to fifty thousand. In addition, there were forty thousand people injured.

There were those Britons who were so critical of Harris that they lost sight of Hitler's frequently expressed desire during the war to see London "burn from one end to the other," while other cities of Britain were pounded to rubble by relentless Luftwaffe bombing. Hitler wrote in *Mein Kampf* that "strength lies not in defense but in attack." That well could have been a maxim stolen from an Arthur "Bomber" Harris textbook on war and how it is won. It is interesting that the anti–Arthur Harris movement that bloomed in the mid-forties among some Britons is alive and well today, coming into its own whenever the name of Marshal of the Air Force Sir Arthur T. Harris is mentioned in print. Although Sir Arthur died in 1984 at ninety-one, passions were aroused in 1992 among British and German peace protesters when they demonstrated against the Queen Mother's unveiling of a statue to the architect of the Bomber Command's "saturation efforts." The Queen Mother, at age ninety-three in 1993, "stepped into another row over Britain's wartime bombing of Germany," reported the London *Sunday Times* in January, 1993. Under the heading "Bomber Harris Exhibition Set to Stir New Rumpus," the story told of an official exhibition marking the seventy-fifth anniversary of the Royal Air Force. The exhibition was staged by the RAF Museum at Hendon in northwest London and was billed in advance publicity as "showing the devastating effects of bombing on the civilian population."

Reaction from British peace campaigners was sharp. Canon Paul Oestreicher, of Coventry Cathedral, which was rebuilt after it was destroyed by German bombing, said: "I think this is very sad. It is a very British hankering for those days of glory, perhaps because we have so little to glory about today. People will wonder why the royals go on to give approval of what seems to many to be a celebration of the bombing of Germany." Supporters of the late Bomber Command strategist countered with leaflets that praised the controversial RAF hero: "Morality? The great *immorality* would have been for us to lose the war that Nazi Germany started. Thank God for 'Bomber' Harris for taking direct action in the only way to shorten the war and save the world from tyranny."

One leaflet was headlined "This Man *Also* Hated Arthur 'Bomber' Harris" and was distributed at Hyde Park Corner to persons listening to a soapbox orator who was on his second day of tirade against Harris. The text of the leaflet read: "Brutality, cold cynicism and an undiluted lust for murder are his chief characteristics. You have only to look into his eyes to know what to expect from such a man. He has the icy-cold eyes of a born murderer. He has accepted a task that many others have declined—the total war against the Huns, as they call us." The leaflet carried the imprint of Joseph Goebbels, Hitler's Minister of Propaganda, who shared responsibility for the killing of six million Jews in Nazi gas chambers. The quote was taken from the Goebbels diaries.

A total of 202 Royal Air Force student pilots (180) and officers (22) were killed in training accidents or died from other causes; their grave sites are at 23 locations in the United States. Carefully tended for more than half a century are the following sites: Montgomery, Alabama (78); Mesa, Arizona (23); Burbank, California (1); Lancaster, California (1); Sacramento, California (4); San Gabriel, California (1); Santa Barbara, California (3); Arcadia, Florida (23); Jacksonville, Florida (2); Warrington, Florida (4); Albany, Georgia (7); Americus, Georgia (1); Trenton, Michigan (3); St. Louis, Missouri (1); Columbus, Ohio (1); Miami, Oklahoma (15); Ponca City, Oklahoma (7); Tulsa, Oklahoma (1); New Castle, Pennsylvania (1); Camden, South Carolina (1); El Paso, Texas (1); San Antonio, Texas (20); and Arlington, Virginia (3).

Memories of more than fifty years are revived each Memorial Day in a plot of ground in Arcadia, Florida, that is forever foreign. A narrow sand road at Oak Ridge Cemetery winds under large oak trees to a neat plot in a back corner. There are 23 identical grave markers in two orderly rows, 12 in the east row and 11 in the west row. Here are buried 23 British cadets who lost their lives during flight training in Arcadia or Clewiston during a five-year period. The first death occurred on July 22, 1941, when nineteen-year-old Charles Russell of Dublin, Ireland, fell victim to meningitis at Arcadia's Carlstrom Field. Russell had arrived at Arcadia on June 9 with the first contingent of RAF students.

With the death of Russell, British authorities arranged with Paul Speer, Arcadia's city manager and recorder, for a burial site at Oak Ridge. The only other death of a Carlstrom cadet was that of Alfred T. Lloyd of Randor, Wales, in a tragic event that never should have happened. Whenever a cadet soloed for the first time, it was customary for his classmates to "initiate" him by dunking him in the swimming pool. Lloyd received warm congratulations from his instructor on Sunday afternoon, January 4, 1942, after the cadet had taken off alone in his Stearman and, after flying over Arcadia, brought his aircraft down in a picture-perfect landing. Four of his friends decided Sunday night that Lloyd should be recognized for his solo achievement; they slipped over to Lloyd's bed, carried him outside and tossed him into the swimming pool. When he went under the water and did not surface, the cadets pulled him from the pool, but were unable to revive him. He was pronounced dead from drowning at 10:00 P.M., seven hours after his solo flight.

Beside the two Carlstrom nonflying deaths, there are two grave sites of Riddle Field cadets who also died when not flying: William Meekin, Lancashire, England, died of meningitis on June 30, 1942, at Miami's Memorial Hospital. (Meekin and Russell were the only cadets who suffered disease-related deaths.) The other cadet who died of a nonflying cause was Louis Wells, York, England, who was killed on December 9, 1941, when he lost control of a car and crashed into a ditch near La Belle, Florida.

The final deaths fell on May 4, 1945, just four days before V-E Day, when the surrender of Germany was announced. Horace Bowley-Booth, of Stafford, England, and Thomas Calderhead, of Fife, Scotland, were killed when their AT-6 crashed at Belle Glade, Florida.

The other RAF cadets (all flying-related accidents) buried at Arcadia are:

East Row
Anthony J. Oakley, Surrey, England, January 14, 1944.
Thomas J. Perry, Glasmorgan, Wales, January 14, 1944.
Ronald A. Purrett, Slough Bucks, England, December 12, 1942.
Geoffrey R. King, Surrey, England, December 3, 1942.
Richard B. Thorp, Staffordshire, England, July 16, 1942.
Marvin H. E. Thomas, Bath, England, April 28, 1943.
Dennis H. Washer, London, England, April 28, 1943.
Edward C. F. Vosper, Devon, England, August 24, 1943.
Leonard G. Stone, Leicester, England, August 24, 1943.

West Row
Lionel M. Viggers, Kent, England, October 4, 1944.
George H. Wilson, Altershoot, Scotland, September 15, 1943.
Robert A. Wood, Middlesex, England, September 15, 1943

Roger B. Crosskey, Bromyard, England, January 20, 1942.
Derek R. Clandillion, Essex, England, January 19, 1943.
John A. Clay, Middlesex, England, January 19, 1943.
Forbes McKenzie Robertson, Essex, England, April 24, 1943.
Michael K. Hinds, Newcastle-Upon-Tyne, England, July 13, 1944.

It should be mentioned that the Arnold Plan cadets at Carlstrom Field were there for only a short period, training on PT-17s before being transferred to basic and advanced at Army Air Force fields. (See Carlstrom Field, p. 5). The British flying training school (5BFTS) at Clewiston had six-month Courses for primary through advanced, thereby facing a much larger risk of accident, based on the number of flying hours and the use of "hotter" aircraft—for example, the advanced AT-6 versus the primary PT-17.

When World War II ended, the United States government asked parents of the deceased RAF cadets if they wanted the remains returned to their respective homes for burial. All parents, or nearest of kin, of those buried at Arcadia chose that the bodies continue to lie where they had fallen in the service of their country.

Many family members of the cadets have visited Arcadia's Oak Ridge Cemetery over the years and have expressed deep gratitude to the citizens of the community and particularly to the K-Post 11 American Legion Auxiliary and Arcadia Rotary Club for the care of the grave sites and the annual Memorial Day service. The simple tombstones carry the Royal Air Force crest above the name of each cadet and date of death. Families submitted expressions of sorrow that appear on the stones in loving memory of young men who had been "Britain's Finest" in peace and war:

> *To the world he was only one. To us he was all the world, sadly missed.*
> *The dearly loved son and brother of a Scottish home.*
> *Let this be my parting word that what I have seen is unsurpassable . . .*
>   *Tagore.*
> *Be ye also ready, for in such an hour as ye think not the son of man*
>   *cometh.*
> *Peacefully sleeping, free from pain, in God's own time we shall meet*
>   *again.*
> *Farewell 'til we meet . . . dads and mums.*
> *Our beloved boy and singing still doest soar, and soaring ever singest.*

One crowded hour of glorious life.

In loving memory of our own son who gave his life that we may live in freedom.

We have loved him in life and will not forget him in death.

Your memory lives in the hearts you loved, eternal rest . . . good night my son.

Father, in thy gracious keeping leave we now thy servant sleeping.

To live in hearts we leave behind is not to die.

Precious only son, the beloved of the Lord shall dwell in safety.

Deep in our hearts his memory is kept . . . we who loved him will never forget.

Deeply mourned by mom and dad.

A proud memory of our dear son. God grant his sacrifice be not in vain.

The spirit of his soul will ever abide with those by whom he was so beloved.

Always in our hearts, Tony dear, we shall meet again my beloved son.

His smiling face no more we'll see, but his memory will live for aye.

To think we were not near to keep vigil over thy bed.

Appendix: Watch Your Language

To the English traveler in America, the language he hears spoken about him is at once a puzzle and a surprise. It is his own, yet not his own. It seems to him a caricature of English, a phantom speech, ghostly yet familiar, such as he might hear in a land of dreams.

Charles Whibley, *The Bookman*, 1908

British cadets at Carlstrom Field could barely restrain themselves from laughter in ground school classes taught by a native of Balmer, Merlin [Baltimore, Maryland], northeast of the U.S. capital, Warshin [Washington, D.C.]. The young Brits tried with varying degrees of success to mimic—but only among themselves—the popular instructor's accent, which today baffles visitors at his Chesapeake Bay seafood lover's paradise. "Back when I went to hoskull [high school] we used to have lunch at a druckstewer [drug store] lunch counter," the instructor would say. "I liked to have a cole race beef sanrich [cold roast beef sandwich]." Totally unaware of the mirth that his accent provoked, the instructor rose to the bait when a cadet asked him how he kept his shirts so white and neat. "Well, my wife always warshes my clothes, wrenches them by hand, and then arns them on an arnin' board." Always a stickler for safety procedures and neatness, the instructor cautioned the British cadets to pay attention to the electric parr [power]. "In case we get any visiting torsts [tourists] on the base, always keep yourself looking neat," he said.

The hilarity brought about over jarring dialects cut both ways. American cadets, not nearly as subtle as the British, would hoot as an RAF physical education instructor at Riddle Field joined them over refreshments at the canteen. A lifelong resident of Stoke-on-Trent, the heart of the potteries country, the instructor had been in the RAF for ten years and had acquired something of a generic accent. But sometimes he would lapse into the speech patterns of his youth. "He's livin' on burred tarm" [borrowed time], he'd state when a cadet ignored flying safety rules. "Watch thee fate." [Watch your feet.] "The dine none room had a teabul an' fower cheers." [The dining room had a table and four chairs.]

Differences between American English and the King's English frequently touched off incidents that ranged from comical to painful because of simple misunderstanding in the use of a word or phrase. When told to report somewhere "directly," a British cadet took it to mean immediately. To an American cadet of southern heritage, "directly" would mean "in a little while." The British cadet wouldn't say, "I'll catch up with you at the mess hall." He would "catch you up at the mess hall."

Many a cadet at Carlstrom or Riddle—Yank or Brit—found himself thoroughly embarrassed at best or severely chastised at worst over the use of a word deemed perfectly proper in its own hometown, but perfectly dreadful in strange surroundings. The word "homely" in the United States generally means unattractive, crude, or downright ugly. In England it can be a compliment, meaning friendly and folksy. Pity the hapless Yorkshire cadet who told his Sarasota date that he found her lawyer father and socialite mother to be a "homely couple." "Fanny," as in "I'll paddle that kid's fanny" is a harmless, if inelegant, American euphemism that means nothing more than buttocks. Fifty years ago, the word "fanny" was not the almost archaic vulgarism that it is today in Britain. But then or now it means only one thing in the UK: the pudendum. It was bad form indeed when a Yank cadet suggested how a Brit could "make progress" with a Palmdale girl he was dating. "All you gotta do, Jack, is give her a little pinch on the fanny when you kiss her."

"Yes thank you, Jeremy," said a sweet young Sarasota girl to her RAF cadet dinner date at the Lido Beach Club. "I'd love dessert." She reached over and clasped his hand. "I tell you, honey, I skipped lunch today and now I really want to get stuffed." [Dig into the chow, tie on the feedbag, take an extra helping].

"Stuffed," Jeremy heard her say in plain Sarasota High School Class of '43 English, bringing joy to his raging 20-year-old hormones. "Stuffed" meant only one thing in Jeremy's plain Manchester English. What a wonderful girl, he thought. Comes right out with it, no need to chivvy [press] if you want to roger her [have sexual intercourse]. He pushed aside his half-consumed dinner.

"Sally," he blurted out, "let's pop over to my room at the John Ringling. We can, you know, do it and I'll get you home before your mum starts to worry."

Prang! (crash). Jeremy's hopes for his first nonprofessional "event" evaporated (three times he had visited, at two dollars a visit, a brothel above the Purity Ice Cream parlor on Main Street). Sally didn't look back as she flew out the door, escaping the clutches of a fiend who had been transformed from sensitive Jeremy Stewart-Browne into Jack-the-Ripper. And it all happened right after soup, gammon, and jacket potato and just before gateau.

American Cadet Brent Hotchkiss stuck out his hand in greeting and said, "Put it there, pal," when he saw a new RAF arrival sweltering in woolen service blues under a relentless 92 degree Arcadia sun. The wilted Brit stared at the outstretched hand and muttered something about thank you for taking "it" as he handed over his heavy duffel bag to the nonplussed Yank.

RAF cadet John Morrow was having coffee at five Sunday morning at the all-night Friendly Cafe ("We Threw the Key Away") when Chuck Liscomb, a popular American advanced instructor and all-round champion skirt-chaser came in, looking like the bitter end of a long, wet night. Concerned over the ashen near-death appearance of one of his favorite instructors, Cadet Morrow nodded at Liscomb, cleared his throat, and felt his face flush as he spoke. "I say, Mister Liscomb, forgive me for saying that you look a bit queer [unwell] this morning."

Liscomb signaled the counterman for coffee, glanced in the mirror behind the counter, and muttered, "Yeah, I know, kid, people take ya for a fruit every time ya wear a fuckin' red bow tie."

RAF Cadet: Mr. Jones is a wizard [excellent instructor]. I hope he gets a good screw [salary].

U.S. Cadet: Hey, don't worry about that guy. He's in like [Errol] Flynn with half the broads on this base.

British cadet Arthur L. Prandle, a newspaperman in civilian life, wrote a weekly column in the *Arcadian*, the weekly newspaper of Arcadia, in which he offered sprightly comments on Carlstrom Field activities as well as touching on topics of interest in the town. His column of June 26, 1941, was headlined "We Speak English—Or Do We? Arcadians and Britons Exchange Talk." Prandle wrote:

Before we British boys out at Carlstrom Field leave for our basic training a subtle change will have taken place in our vocabularies and in those of our friends in Arcadia, whose number increases each week.

While we love to listen to you and to your expressions—many of which sound extremely quaint to us, you evidently just love to hear us speaking what we have always been led to believe is English.

Dale Delanty, one of our popular American instructors, has already been musing on the probable effect that the Florida accent will have upon us when we return home. "I say, old boy, lookie down yonder a piece," is Delanty's prediction.

We won't be the only ones to undergo a change in speech patterns. One of Arcadia's most charming ladies remarked over the weekend that "petrol" was quite dear in England: in more than one case I have noticed a gradual influx of English expressions creeping into the conversations of my American friends.

Yes, when it's our time to leave this pleasant spot, I can see us begging English lessons from the townsfolk and in return teaching them to speak once again like real Florida crackers.

✧ Bibliography ✧

Amory, Cleveland. *The Last Resorts*. Harper & Bros., 1948.

Ash, Jennifer. *Private Palm Beach*. Abbeville Press, 1992.

Baker, Newton D. *Frontiers of Freedom*. George H. Doran Co., 1918.

Birmingham, Stephen. *The Grandes Dames*. Simon & Schuster, 1982.

Blumenson, Martin. *The Duel for France 1944*. Houghton Mifflin, 1963.

Brown, Warren, editor. *Newsletter of the Florida Aviation Historical Society*. Indian Rocks Beach, Florida.

Bryson, Bill. *Mother Tongue: English & How It Got That Way*. William Morrow & Co., 1990.

Churchill, Winston S. *Their Finest Hour*. Houghton Mifflin, 1949.

Craven, W. F.; J. L. Cate. *The Army Air Forces in World War II: Volume III: Argument to V-E Day*. Cambridge University Press, 1951.

Curl, Donald W. *Mizner's Florida American Resort Architecture*. MIT Press, 1992.

deQuesada, A. M. *The Royal Air Force over Florida*. Arcadia Publishing ("Images of America" series), 1998.

Douglas, Marjory Stoneman. *The Everglades: River of Grass*. Pineapple Press (revised), 1988.

Dundas, Hugh. *Flying Start*. Penguin Books, 1990.

Francis, Devon. *Flak Bait*. Duell, Sloan and Pearce, 1948.

Guinn, Gilbert S. "British Aircrew Training in the United States, 1941–1945," *Journal of Southwest Georgia History*, 59–80, Albany State College, 1992.

Harkin, Jeremy. *The Military Aircraft Archive*. Internet site: http://www.militaryaviation.com.

Hillary, Richard. *The Last Enemy*. Ulverscroft, 1943.

Homan, Lynn M. *Wings over Florida*. Arcadia Publishing ("Images of America" series), 1999.

Johnston, Alva. *The Legendary Mizners*. Farrar, Strauss and Young, 1953.

Keegan, John. *The Second World War*. Penguin Books, 1990.

LaHurd, Jeff. *Sarasota: A Sentimental Journey*. Sarasota Alliance for Historic Preservation, 1991.

———. *Sarasota Then and Now*. Sarasota Alliance for Historic Preservation, 1994.

McCollister, John; Diann Ramsden. *The Sky Is Home: The Story of Embry-Riddle Flying School.* Jonathan David Publishers, 1986.

McCrum, Robert; William Cran; Robert MacNeil. *The Story of English.* Viking Penguin, 1986.

McCullough, David G., ed. *American Heritage Picture History of World War II.* Bonanza Books, 1966.

McKenney, Ruth; Richard Bransten. *Here's England.* Harper and Row, 1955.

Mencken, H. L. *The American Language* (fourth edition). Alfred A. Knopf, Inc., 1936.

Moss, Norman. *British/American Language Dictionary: For More Effective Communication Between Americans and Britons.* Passport Books, 1988.

Nicolson, Harold. *The War Years: Diaries and Letters 1939–1945.* Atheneum, 1967.

Pyle, Ernie. *Brave Men.* Grosset & Dunlap, 1944.

Reston, James. *Deadline: A Memoir.* Random House, 1991.

Seward, Dudley. *Bomber Harris: Marshal of the Royal Air Force.* Doubleday, 1985.

Shirer, William L. *20th-Century Journey: The Nightmare Years, 1930–1940.* Little, Brown & Company, 1984.

Smalley, Jonathan. *The V-J Hangover.* The Book Guild Ltd., 1987.

Terraine, John. *The RAF at War.* Time-Life Books, 1981.

———. *A Time for Courage: The Royal Air Force in the European War.* Macmillan, 1985.

Turner, E. S. *The Phoney War.* Michael Joseph Ltd., 1961.

White, Theodore. *In Search of History.* Harper & Row, 1978.

∾ Index ∾

258 ∾ *Index*